Quality and innovation in food chains

Lessons and insights from Africa

Quality and innovation in food chains

Lessons and insights from Africa

edited by:
J. Bijman
V. Bitzer

Wageningen Academic
P u b l i s h e r s

EAN: 9789086862801
e-EAN: 9789086868254
ISBN: 978-90-8686-280-1
eISBN: 978-90-8686-825-4
DOI: 10.3920/978-90-8686-825-4

First published, 2016

© Wageningen Academic Publishers
The Netherlands, 2016

Wageningen Academic Publishers,
P.O. Box 220, 6700 AE Wageningen,
The Netherlands,
www.WageningenAcademic.com
copyright@WageningenAcademic.com

Table of contents

Preface

This book is one of the outputs of the 'Co-Innovation for Quality in African Food Chains' (CoQA) programme. CoQA was an international and interdisciplinary research and capacity building programme with the aim of providing insights that can help to improve quality in food chains in Benin, Ethiopia and South Africa. The CoQA programme was administered by Wageningen University, and run from 2008 until 2016.

The CoQA programme was a collaboration among Wageningen University in The Netherlands, University of Fort Hare in South Africa, Hawassa University in Ethiopia, and Université d'Abomey-Calavi in Benin. We would like to thank all partners for their sustainable commitment to the CoQA programme.

We owe much gratitude to the Interdisciplinary Research and Education Fund (INREF) of Wageningen University which has funded the main part of the CoQA programme. INREF aims at promoting interdisciplinary and participatory research approaches, strengthening international research and development partnerships, and building capacity in developing countries. The results reported in this book all arrive from interdisciplinary, international and capacity developing partnerships.

We hope this book is inspiring for all researchers, policy makers, and practitioners involved in studying, supporting and implementing quality improvements and innovations in food chains. The chapters in this book proof that interdisciplinary research leads to better understanding of the pathways for quality upgrading. Finally, our co-innovation approach has shown that only a joint effort of public and private actors will lead to sustainable improvements in food quality.

Jos Bijman, project manager CoQA
Verena Bitzer, post-doc researcher CoQA

Wageningen, March 2016

1. Quality improvement in food value chains: searching for integrated solutions

J. Bijman[1][*] and V. Bitzer[2]

[1]*Wageningen University, Management Studies Group, P.O. Box 8130, 6700 EW Wageningen, the Netherlands;* [2]*Royal Tropical Institute (KIT), P.O. Box 95001, 1090 HA Amsterdam, the Netherlands; jos.bijman@wur.nl*

Abstract

Quality improvement in food value chains offers both opportunities and challenges for farmers in Africa. This chapter introduces the key concepts that are used in the studies presented in this book. It also provides a short description of each of the chapters. Quality is an elusive concept. It has a different meaning for each of the different value chain actors involved in producing, processing, trading and consuming food products. Some of these quality preferences can easily be measured, others are much more difficult to detect. This has implications for monitoring and control, such as in quality assurance systems, but also for providing proper economic incentives for each of the value chain actors. Finally, it has implications for the alignment of quality preferences throughout the value chain. The latter is important because the opportunities for quality improvement can only be understood by analysing the chain as a whole and assessing the motives and capabilities of all chain actors. In this chapter we also explain the interdisciplinary perspective we take on studying quality improvement and innovation. As quality improvement is a type of innovation process, the literature on innovation processes and innovation systems can be used for better understanding the options and constraints for quality upgrading in food chains in developing countries. Based on the recognition that innovation processes involve multiple actors, at multiple levels and engaged in multiple activities, at the end of the chapter we present the co-innovation.

Keywords: quality, value chain, innovation, Africa, interdisciplinarity, co-innovation

1.1 Introduction

Quality improvement in food chains offers both opportunities and challenges for farmers in Africa. In recent decades, the production, distribution and consumption of food has changed dramatically, particularly in Sub-Saharan Africa. The demand for higher quality of food products is both one the causes and one of the consequences of the major changes. Due to globalisation of food systems, enabled by new technologies and reduction of trade barriers, farmers in Africa have found new market opportunities in producing for developed country markets. Also the rise of domestic supermarkets, catering for a growing population of middle income consumers, provides new and potentially more remunerative outlets for farmers. These high end markets have increasingly strict requirements in terms of quality, certification, traceability, product uniformity, delivery times and food safety (Reardon *et al.*, 2009). As not all farmers have the resources and skills to produce for the highest quality markets, an increasing differentiation can be seen in the development of food chains, mainly determined by the quality demands in the final consumer markets that these products are targeting.

Quality, however, is not the only concern of the agricultural industry and those supporting this industry. Also quantity is an issue, as obtaining food security for a rapidly growing population, particularly in Africa, is not self-evident. In order to improve food security, the productivity of smallholder farms needs to increase. To induce smallholders to increase production, a combination of pull and push factors needs to be in place. Agricultural support programmes of national governments, international organisations, donor organisations and non-governmental organisations (NGOs) seek to strengthen market access, as the key mechanism to obtain both higher production and better quality.

The main assumption underlying the developmental effort of national and international organisations is that enhanced market access will provide smallholder farmers with the capabilities and incentives to both increase production and produce the right products. Increasing production should come from better access to inputs, credit and technical assistance. Producing the right products implies a better exchange of information between producers and consumers. Being able to produce more and better products is directly depending on access to the inputs and services needed on the farm. Thus, access to input markets and access to output markets are strongly interdependent.

While acknowledging the importance of food security, this book will focus on the improvement of food quality: how can smallholder farms improve the quality of their products and how can this quality be maintained or even upgraded throughout the food chain? Improving quality will lead to higher food chain income, higher consumer satisfaction, and lower environmental impact (as food losses will be reduced).

In answering the question how food quality can be improved, this book adopts an *interdisciplinary perspective*. Improving quality is not just a matter of developing and adoption of new technologies. While technical innovation in cultivation, logistics or processing may generate higher quality and reduce quality degradation, implementing such innovations requires complementary innovations in organisation and institutions. Therefore, this book seeks to explore integrated quality solutions, which combine changes in technology with changes in the food chain organisation and changes in the institutions that affect the decisions of individuals and organisations in the food chain.

In search of integrated quality solutions, this book presents a collection of case studies based on field research in various African food chains. By analysing the experiences of different chains, different crops and different countries, the book aims to generate and disseminate practice-oriented knowledge concerning quality and innovation in food chains. The lessons learned and insights obtained from the case studies will help to find practical answers to some of the key challenges that the agricultural sector in Africa is currently faced with. Based on the findings of each chapter, the book will also provide directions for further research.

All of the chapters of the book utilise a *food chain perspective*. A food chain comprises the subsequent activities and actors that are involved in producing, transporting, processing, trading and consuming food products (Trienekens, 2011). Using this perspective implies acknowledging that quality improvement is the result of the coordinated activities of different actors producing and handling the food products. This implies that quality improvement activities of the farmer will only be successful in changing his/her livelihood situation when the objectives of these activities are implicitly or explicitly aligned with the objectives of the other actors in the chain. Thus, our chain perspective covers both the coordination of operational activities, such as logistics and information exchange, and the alignment of the economic interests of each of the chain actors.

Throughout the book, we use an *innovation perspective*. We conceptualise quality improvement as a process of innovation, thus paying attention to issues inherent to all innovation, such as cognitive challenges, entrepreneurial attitudes, market opportunities, competitive pressure, organisational support, institutional barriers and technological advances. More specifically, we use an innovation systems perspective, which means that innovation is always the result of the interplay of many actors and many processes (Hall *et al.*, 2001). In order to be successful, innovation requires collaboration among actors from business, government and NGOs, to jointly generate and implement new knowledge (Sumberg, 2005).

This introductory chapter presents the main concepts that are used in this book. The structure of the chapter is as follows. First, we will discuss the concept of quality and its different definitions. Particularly relevant here is that different actors involved in producing, processing, trading and consuming food products may have diverging demands for the quality of these food products. Second, we will present our food chain perspective, as opportunities for quality improvement can only be understood by analysing the chain as a whole and assessing the motives and capabilities of all chain actors. Third, we will explain our interdisciplinary perspective on quality improvement and innovation. As quality improvement is a process of innovation, the body of knowledge on innovation processes and innovation systems provides a better understanding of the options and constraints for quality upgrading in food chains in developing countries. Based on the recognition that innovation processes involve multiple actors and multiple activities, we develop the concept of co-innovation. Finally, we present a short outline of each chapter of the book.

1.2 Defining food quality

Food quality is an elusive concept as it has different meanings for different actors in the food chain. In the academic literature, a wide range of definitions of quality prevail based on the perspective adopted by various disciplines. Reeves and Bednar (1994) distinguish between four broad approaches to quality: quality as excellence; quality as value; quality as conformance to specifications; and quality as meeting and/ or exceeding customers' expectations. In the marketplace and hence in the context of market access and competitiveness, quality is usually viewed as 'perceived quality' and ultimately depends on the judgement of the customer (Oude Ophuis and Van Trijp, 1995; Reeves and Bednar, 1994). Quality as satisfying customer expectations implies that it is time and location specific (Luning and Marcelis, 2006). While for professionals in handling food products quality means the extent of conformance to specifications, often customers do not know how well a product or service conforms to those specifications (Reeves and Bednar, 1994). This implies that quality includes both subjective aspects, such as consumer preferences, as well as objective aspects, such as the measurable compliance with specific quality requirements (Ruben *et al.*, 2007).

Steenkamp (1989) posits that overall quality judgement is based on intrinsic and extrinsic quality cues. Intrinsic cues refer to the physical product properties, such as colour, shape and size, whereas extrinsic cues are related to the product, but are not physically part of it. Extrinsic cues include, for instance, the brand name and the social and environmental conditions of the production and distribution process, such as the number of food miles (Coley *et al.*, 2009).

The marketing literature has identified three types of perceived quality attributes: search, experience and credence (Darby and Karni, 1973; Nelson, 1970). Search attributes are the quality attributes known at the moment of the purchase; for instance, the apple has no external sign of decay. Experience attributes are known after personal experience; for instance, the apple has a good taste. Finally, credence attributes are quality attributes that cannot be verified through search or experience: for instance, the apple has been produced in an organic production system, thus without the use of any chemical pesticides. In the domain of agriculture and food this distinction into search, experience and credence attributes is extremely relevant, because of the emotional attachment of people to specific foods, the importance of quality for human health[1], and the increasing importance of sustainability in the food chain which leads to credence attributes becoming more important.

From a food chain perspective the various preferences and interpretations of food quality of different actors in the chain need to be aligned. For instance, the quality attributes that might be important to producers, such as yield and disease resistance of a crop variety, may matter little to retailers who are more concerned about the shelf life and appearance of a food product. For a processor the uniformity of products and the

---

[1] In the Codex Alimentarius (FAO/WHO, 2006) food safety is an integral part of food quality.

time of harvest are crucial quality variables for the optimal alignment of production and processing activities. Finally, the consumer may particularly be interested in the healthiness and convenience of a product. Table 1.1 shows the different interpretations of quality by different actors along the food chain. The challenge for the food chain as a whole is to reconcile these different interpretations of quality (Ruben *et al.*, 2007). In other words, chain alignment requires the implementation of information exchange and joint decision-making structures that can deal with the divergent and dynamic preferences of all chain actors.

From a public policy perspective, food quality concerns traditionally focussed on food safety issues (Luning *et al.*, 2006). More recently, also nutritional quality has been recognised as an important public issue. Nutritional quality (or nutrition security) goes beyond availability and safety of food products, as it emphasises both the availability of specific nutrients necessary for particular groups of consumers (such as infants and elderly) and the balance among various food ingredients. For instance, consumers that have more income to spend usually shift their diets from carbohydrates to protein and animal fats, which entails a less healthy consumption pattern (Popkin, 1999). While food safety and nutritional quality are (or should be) basic conditions for food products, this book focusses on those quality issues that lead to differentiation among products and markets and thereby to differentiation among food chains.

1.3 Different perspectives of food chain analysis

In the world of international development, value chain analysis as a research tool and value chain development as an intervention strategy have become popular among NGOs, consultants and applied researchers (Donovan *et al.*, 2015; Trienekens, 2011). Within the academic literature, no consensus exists on the exact definition of a value chain or value chain development. Researchers from management and business, often influenced by the literature on Supply Chain Management and Strategic Management, conceptualise the value chain as a set of activities that need to be aligned in order to get efficient logistic processes, effective coordination between production and distribution, and efficient information exchange, such as through tracking and tracing

Table 1.1. Interpretation of quality by various food chain actors (Ruben et al., 2007: 30).

Actor	Quality aspects
Breeder	vitality of seed, yield
Grower	yield, uniformity, disease resistance
Wholesaler	shelf life, availability, sensitivity to damage
Retailer	shelf life, diversity, exterior, low losses
Consumer	taste, healthiness, perishability, convenience, constant quality

systems. In research on and managing of the food value chains, improving quality and innovation is one of the focus activities (Fritz and Shiefer, 2008; Trienekens, 2011).

A value chain has also been conceptualised as a strategic network (e.g. Lazzarini *et al.*, 2001). Value chains are networks of collaborating but independent business organisations with the joint objective of providing final consumers with good quality products. The emphasis in the strategic network literature is, on the one hand, on seeking ways of collaboration that support joint innovation but maintain individual entrepreneurship, and, on the other hand, on competition among different networks that supply similar products to the final customer. Thus, the idea of value chains competing against each other is in line with this strategic network thinking.

Yet another approach, often taken by both orthodox and political economists, emphasises the value chain as a set of actors. Each of these actors has its own objective function, often profit-driven. For the value chain as a whole, the challenge is to align the interests of all participating actors in such a way that each actor has sufficiently strong incentives to participate and to continue to contribute to total value generation. Value chain analysis, then, focusses on describing the actors involved, calculating the gross margins or even net margins at each stage of the chain, and explaining the contractual linkages that are used for the bilateral transactions among chain actors. The political economy strand of value chain analysis also studies the power relationship among chain actors in order to find out which actor is most influential in moving the whole chain. In other words, it investigates which actor has the most power to determine what quality improvements will be required and enabled.

In the Global Value Chain literature applied to food chains it has been shown that particularly large multinational food companies as well as international supermarket companies are the dominant actors in food chains (Humphrey and Schmitz, 2002). These companies have a strong influence on whether and what type of innovations in the chain will be adopted (Saliola and Zanfei, 2009). The food quality and safety standard GlobalGAP is often presented as an example where large (originally European) supermarket companies have set the private standards that all suppliers of fresh produce have to comply with (Hatanaka *et al.* 2005) The introduction of these standards by large supermarket companies has greatly supported the introduction of sustainable production methods. But it has raised many concerns about the inclusion of particular groups of producers in developing countries that may not (yet) be able to apply the strict requirements due to a lack of resources and capabilities (Lee *et al.*, 2012).

The different conceptualisations of the value chain – whether it is a set of activities, a set of actors or a strategic network – have implications both for the type of value chain analysis and for the design of interventions that follow from the analysis (Donovan *et al.*, 2015). With an activity-based definition, value chain analysis and value chain interventions focus on improving the efficiency of production and distribution processes. This 'engineering' approach may lead to optimisation strategies and innovation policies that clearly show what needs to be done to obtain improvement, but that are often not clear on which actors have to implement what changes. Under the actor-defined value chain approach, a distinction can be made between resource-poor and resource-rich actors, and the link between them is the object of study. If development is about enabling the resource-poor to benefit from new market opportunities, then the goal of value chain development is to help smallholders link up with modern value chains by providing them with the knowledge and skills they need and by strengthening their bargaining power vis-a-vis the other chain actors. The strategic network perspective on value chains leads to the identification of the dominant companies in the network in order to encourage these actors to support their suppliers in improving quality.

1.4 Market access and the debate around the impact of quality standards

1.4.1 Linking farmers to markets

A vibrant debate has emerged over the past decade on 'linking farmers to markets' to make use of the productive capacities of smallholder farmers and promote rural development. Most of this debate has focused on how emerging quality standards facilitate or hinder smallholder market access (Henson and Reardon, 2005; Jaffee *et al.*, 2011). While case studies have provided ample evidence for both scenarios, the prominence of this debate has largely detracted attention from the most fundamental challenge underlying all market access strategies of smallholder farmers: the issue of integrated quality improvement. Quality improvement can entail but exceeds compliance with standards, recognising that quality relates to a broad range of often interrelated issues, including market orientation (national or international), organisation and logistics of the chain, and technical aspects of production, processing and distribution. This type of integrated research on food quality problems has so far been limited to developed countries. The re-emergence of agriculture for development on the policy and research agendas (e.g. World Bank, 2007a) warrants a systematic analysis of integrated quality solutions that can strengthen smallholder market access and competitiveness in national and international food chains.

Market access[2] has been identified as one of the most important factors influencing the performance of smallholder farmers in developing countries. Access to more remunerative markets, such as those for products with higher value, is now considered as a major pathway to enhance and diversify the livelihoods of low-income rural households and thereby reduce poverty in the rural area. Many development NGOs

---

[2] With raising the issue of market access we do not want to suggest that smallholder farmers never sell farm products. Even in the poorest rural areas there are few farmers that are pure subsistence farmers (i.e. producing only for their own consumption). What we mean by a lack of market access is that many farmers do not have access to more remunerative output markets, or do not have access to well-functioning inputs markets. Lack of market access thus prevents smallholders to introduce the new farming technologies (such as new crop varieties) needed for quality improvement.

and donors have, over the last decade, made improved market access as one of their key targets in supporting farming households, and have selected value chain development as the core method for achieving such better market access.

However, market access is still a major problem for many smallholders, resulting from a combination of market characteristics and features of the smallholders. Constrained market access may result from physical barriers such as poor infrastructure, from weak technical capacity of the farm and farmer, from a lack of resources to make the necessary investments for meeting quality demands, or from the inability to bear the risk related to investment in quality improvement (World Bank, 2007b).

Low market access may be caused by problems in production or in transacting with buyers. Production problems often result in the inability of the farm to increase the quality of its products. This could be due to lack of knowledge, lack of scale, lack of resources, lack of access to inputs and technical assistance, lack of labour, and high risk related to local agro-ecological conditions. Also the need to produce sufficient food for farm household consumption may prevent farmers to shift to the crops and varieties that would strengthen their market position.

All transactions involve transactions costs, but these costs may vary substantially depending on the type of market and the type of buyer. Transaction costs are the cost of collecting information (e.g. on supply and demand and on prices), the cost of contracting (e.g. negotiating cost), and the cost of contract enforcement (e.g. ensuring that the buyer will pay and will pay on time). Transaction costs are particularly high for small producers who rely on traders and middlemen to come to the farm to collect the product. Lacking transport means and information about market conditions places these farmers in a disadvantaged bargaining position. Farmer transaction costs can rapidly increase when farmers switch from generic products for local markets to specialty products for export markets.

Switching from generic product quality, which is understood and appreciated by local buyers, to specific product quality demanded in distant markets entails the risk of becoming dependent on the few buyers that supply those distant markets. Such dependency is risky for smallholder farms, particularly because farmers need to invest at the start of the growing season and only receive payment for their products after the harvest (and then even after a delay). During this time lag, many uncertainties may arise. Risk-averse farmers will not easily accept these uncertainties and the risks involved (Chavas, 2004).

1.4.2 Markets, quality requirements and upgrading

Quality requirements differ considerably between markets (Poulton *et al.*, 2006). Local (wet) markets tend to have few quality requirements, mainly related to physical quality attributes, such as appearance. Local supermarkets are still modest in their quality requirements. Foreign supermarkets, particularly those in Northwest Europe and the USA are the most demanding in terms of quality attributes. In between these three alternatives, other levels of quality requirements can be found.

Jaffee *et al.* (2011) distinguish six different markets which represent six different levels of quality requirements. These six markets range from wet markets and small retail stores in the local community of the producer in a developing country all the way to high-end supermarkets in developed countries. The authors pose that moving from one market to the other (from lower to higher quality requirements) entails a step in a process of upgrading (Figure 1.1). The different markets available to smallholder producers represent alternative marketing choices and supply response strategies. The key message of Jaffee *et al.* (2011) is that each upgrading step entails higher levels of quality requirements, which implies higher levels of compliance costs (including on-farm investments) and transaction risks. Going from left to right does not necessarily represent a better income for the farmer, as both revenues and costs go up. Only when farmers are able to produce at a higher quality level while reducing the costs and risks, the next step of upgrading will be attractive.

1.4.3 Private quality grades and standards

In the past, food quality and safety standards were mainly introduced and monitored by public authorities, who had two main reasons to develop and implement quality standards. First, by implementing and particularly by controlling compliance with specific quality requirements, the state can protect consumers from opportunistic producers and traders. This function is particularly relevant for those quality attributes

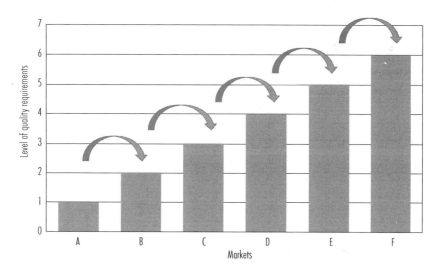

Figure 1. Upgrading options: moving from low quality to high quality markets (adapted from Jaffee et al., 2011).

that affect the health of consumers (notably those related to food safety). Second, quality standards reduce transaction costs by reducing asymmetric information, thereby making markets more efficient. When standards apply, the cost of market participants to find out about the quality of a specific product are reduced, which leads to a general welfare gain. Introducing standards and a system of compliance monitoring reduces transaction costs in two ways: customers spend less time and effort on finding out about the quality of a particular product, and, because they trust the quality grade, they are willing to pay the proper price for this grade. In other words, the adverse selection problem is prevented, and the producer receives a proper incentive to maintain the quality of the product.

While in the past food quality and safety standards where mainly a responsibility of the state, the last two decades have seen a proliferation of private standards (Hatanaka *et al.*, 2005; Henson and Humphrey, 2010). The introduction of private standards has important implications for the governance of food chains and the inclusion of smallholder farmers. The effects of private standards, however, are not straightforward, and a major debate has appeared in the development studies literature on the impact of private standards on developing countries in general and on smallholder market access in particular.

Why have private companies introduced quality grades and standards that complement or even replaced public standards? Henson and Reardon (2005) discuss three reasons for the introduction of private food quality standards. First, in the case of inadequate public regulation and enforcement, private companies may feel urged to set up their own system of standards, particularly to protect their reputation. As marketing costs make up an increasing part of their total costs, food companies are keen to assure the quality of their products towards consumers. Second, companies may consider public regulations not sufficiently strict, particularly when consumer demands change, and therefore introduce their own standards. Thirdly, companies use private standards in their differentiation strategy. For instance, the first food companies (producers and retailers) that offered fair trade products could position themselves as being more concerned with social responsibility than their competitors.

As the introduction and conformity assessment of standards entails a costly investment, most private standards are not introduced by individual companies but by groups of companies. For instance, the introduction in the late 1990s of EurepGAP (now GlobalGAP) can be explained by the need of major supermarket companies to impose the same quality standards on all their suppliers (Van der Grijp *et al.*, 2005). By making sure that all suppliers comply to the same quality standards, dependency of the buyer on one supplier is prevented and competition among suppliers strengthened. At the same time, the introduction of EurepGAP/GlobalGAP gave supermarkets the lead in pushing for more sustainable production methods, particularly for fresh produce. Developing this system of standards was also a pre-emptive strategy against those suppliers that were trying to set up their own set of standards. Finally, introducing a set of quality standards fits in the strategy of most supermarkets to strengthen the competitive position of their private label (in-house brand) products. In response to retailers setting up their private quality standards, also agricultural producers and food processing companies have developed their own quality standards.

Another reason for companies to introduce private quality standards is the need for risk management (Henson and Humphrey, 2010). On the one hand, changes in European food law have placed a larger responsibility on the private sector. Companies in the food chain need to show that they have a food safety control system in place, not only for their own operations, but also for the supplies that they purchase. On the other hand, the globalisation of food chains has created new sources of risk as food is subject to more transportation and handling activities. While compliance with public food quality and safety standards is mainly measured within the jurisdiction of the public authorities, food companies operating in global food chains need to control their supply chains beyond the national borders. Private standards, in combination with third-party compliance monitoring, can ease this task of controlling global food chains.

Henson and Humphrey (2009) present a typology of private food quality and safety standards. They distinguish between individual company standards, such as Tesco's Nature's Choice and Carrefour's Filières Qualité, collective national standards, such as the Farm Assured British Beef and Lamb and the QC Emilia Romagna (Italy), and collective international standards, such as GlobalGAP and Marine Stewardship Council. While the initiative for setting up private collective standards was taken by a group of processing or retailing companies, they are often discussed in multi-stakeholder platforms where producers, traders, consumers and NGOs are now jointly deciding on the quality requirements.

1.5 From innovation to co-innovation

In this book, quality improvement will be conceptualised as a form of innovation. Improving the quality of products or production processes, changing the quality control system, or even revising the governance structure used for the transactions in the value chain are all elements of an innovation process that has the objective of delivering better food products to consumers.

Innovation is understood as 'an on-going process of learning, searching and exploring, which result in new products, new techniques, new forms of organisation and new markets' (Lundvall, 2010: 8-9). Innovation does not have to refer to a worldwide novelty, but rather to something perceived as new in a particular locality or by particular actors. Especially in a developing country context, the adoption and adaptation of existing knowledge or technologies are most relevant (Aubert, 2005; Pietrobelli and Rabellotti, 2011).

Over the past few decades, thinking on innovation has altered significantly, shifting from the dominance of linear perspectives on technological change towards the wide acceptance of the innovation systems (IS) perspective. An innovation system is defined as 'a network of organisations, enterprises, and individuals focused on bringing new products, new processes, and new forms of organisation into economic use, together with the institutions and policies that affect their behaviour and performance' (World Bank, 2007b: 18). The IS approach recognises that actors do not innovate in isolation, but depend on extensive interaction with their environment (Fagerberg, 2005).

This book starts from the premise that innovation cannot be understood from a technical perspective alone, as innovations form part of a 'successful combination of hardware, software and orgware' (Smits, 2002: 865). Hardware relates to appropriate technologies, software to knowledge and mind-sets, and orgware to organisational and institutional conditions that influence the development and diffusion of innovations. Such an integral view on innovation recognises the complementarities between technical, organisational and institutional dimensions, and places attention to the interaction among the various stakeholders involved. In the case of food value chains, innovation involves all actors that together form the chain.

The integral perspective leads to the concept of co-innovation, defined as complementary innovations combining different dimensions (technological, organisational and institutional changes) for integrated quality improvement at different levels of the value chain and involving different actors. More specifically, co-innovation has been defined as the combination of collaborative, complementary and coordinated innovation (Bitzer and Bijman, 2015). Collaborative refers to the multi-actor character of the innovation process, where each actor brings in specific knowledge and resources. Complementary indicates the smart combination of technological, organisational and institutional innovation. Coordinated draws attention to the importance of chain-wide adjustments and changes to make innovation in one stage of the chain a success. Most chapters of this book use the co-innovation perspective in exploring and evaluating the different options for quality improvement in African food chains.

Understanding the options for quality improvements as an co-innovation process requires an interdisciplinary perspective in research. 'Interdisciplinarity is defined as collaborative work between scientists, each from different disciplines, each with its own concepts, methods and epistemology, working together on the same research question, mutually influencing each other and needing some shared concepts and methodologies' (INREF, 2010). Interdisciplinary research is more challenging than monodisciplinary studies, as it require learning the 'language' and methods of other disciplines. However, in the end, it may lead to a better understanding of real life problems and generate better solutions, because explanations and solutions from different disciplines are being assessed in an integrated framework.

As quality improvement involves both changes in technology (such as new crop varieties, new sorting machines or new storage facilities) and change in organisation and institutions (such as new contract forms or new quality standards), only an interdisciplinary perspective can generate the full explanation of the opportunities chain actors may have. Finding the optimal mix of changes given the local context, and implementing the various changes simultaneously requires an analytical framework that integrates different disciplines, ranging from crop sciences, to food technology, to economics, and to other behavioural sciences. Most of the chapters in this book have taken an interdisciplinary perspective, combining insights from social sciences and technical sciences to obtain better explanation and to generate recommendations for improvement.

1.6 Outline of the book

The book deals with quality improvements and innovation in various African food chains. In the original research programme 'Co-innovation for Quality in African Food Chains', three countries and three products were chosen: the pineapple value chain in Benin, the potato value chain in Ethiopia, and the tree fruit value chain in South Africa. In the latter case, the programme studied both citrus and deciduous fruit. Each of these value chains is covered in two chapters. In addition, three chapters on food chains in other African countries have been added, because these cases also give good examples of interdisciplinary co-innovation processes towards quality improvement.

Preceding the eight chapters with cases studies, Chapter 2 discusses the different organisational and institutional arrangements that can be found in value chains. Such arrangements are a reflection of the extent of collaboration among the chain partners, but also of the power relations in the chain. The chapter focuses on three often found organisational and institutional arrangements: contract farming arrangements; producer organisations and public private partnerships. While each of these arrangements affects the options for quality improvement individually, they can also be found in combinations.

Chapter 3 explores the key quality issues of the Beninese pineapple sector. As pineapple is being commercialised in three different value chains – for local markets, for neighbouring countries and for Europe – also quality requirements differ. An analysis is made of the constraints and opportunities for improving pineapple quality at production, processing, trading and post-harvest levels. The chapter concludes with recommendations on how various actors in the institutional environment of Benin, such as governmental agencies and NGOs, can contribute to quality improvements of the pineapple sector.

Access to up-to-date information on market prices and quality requirements remains a key issue for smallholder farmers' access to high income markets. The aim of Chapter 4 is to explore the problem of information asymmetry between farmers and buyers in the pineapple supply chain in Benin, and to assess strategies using mobile phones to overcome this problem.

Chapter 5 focusses on the seed potato chain in Ethiopia, distinguishing among three different seed potato systems: the informal system, the alternative system and the formal system. The chapter analyses the performance of seed potato value chains with respect to their ability to supply quality seed tubers to seed potato systems, by using the chain performance drivers enabling environment, technology, market structure, chain coordination, farm management, and inputs.

Chapter 6 deals with ware potato chain. It analyses the extent of quality alignment along the chain, focussing on the varieties that ware potato growers choose. There seems to be a large misalignment between the new potato varieties that have become available from breeding institutes and that have better agronomic quality and the varieties that wholesalers and retailers downstream in the value chain ask for. Thus, ware potato farmers continue to grow traditional varieties that have suboptimal agronomic quality but have high quality for the final consumer.

The recent turn towards the use of private quality standards in global agrifood chains has triggered an intense debate among scholars about the implications for smallholder producers in developing countries wishing to access such chains. Chapter 7 deals with the citrus chain in South Africa, particularly with the question whether smallholders have a chance to participate in export-lead high quality value chains. As the South African government has introduced a number of new institutional arrangements (IAs) between smallholder farmers and established agribusinesses, the chapter discusses these arrangements and the contribution they make to quality improvement and smallholder inclusion.

While Chapter 7 deals with the collaborative arrangements in the value chain needed for smallholder market access, Chapter 8 explores the farm resources that smallholder producers require to participate in high value markets. Using case studies from the Western Cape Province of South Africa, the authors identify resources that smallholder producers in developing countries require to increase competitiveness and sustain participation in high value markets.

Chapter 9 deals with the important role the institutional innovation plays in making quality improvement possible and feasible. The focus is on rural collective action that enables small farmers to participate in newly created export chains. The chapter compares two cases: new apicultural technologies in the North West of Uganda and high value horticulture exports crops in the North of Peru. This chapter seeks to unravel which factors and actors play what roles and how these explain differences in the process of institutional development and in that way to arrive at a better understanding of local institutional change.

Also Chapter 10 deals with how the institutional environment affects the options for quality improvement and market access for smallholder producers in African countries. This chapter looks at the export of shrimp from Benin. Stable market access for shrimps is hindered by the microbiological and chemical characteristics that affect product quality and safety. In the international market, these quality aspects have legal implications, potentially leading to import bans if safety standards are not met. This chapter examines the quality and legal issues of the Beninese shrimp chain and discusses the responsiveness of the chain to these issues.

High variation in quality is one of the limiting factors for market access, particularly for export markets. Chapter 11 presents a case study of quality measurement of a non-timber forest product, arabic gum. The chapter explores the possibility to understand the current practices of producers in terms of quality supply and to link at least some of the users' quality criteria to production and marketing practices of producers. The study finds that good quality as defined in the field is not always good when measured in laboratory; yet improving quality in the field increases the likelihood of obtaining chemically good gum. Furthermore, determinants of supply by collectors and traders are investigated for two quality attributes namely size and cleanliness of gum nodules. Quality maintenance and improvement is influenced by harvest and post-harvest practices, behaviour and experience of traders, and price expectations.

The final chapter, Chapter 12, presents a discussion of the main findings of the various chapters. It follows the main concepts as introduced in Chapter 1, that is, food chains, interdisciplinary research, and co-innovation.

References

Aubert, J.-E., 2005. Promoting innovation in developing countries: a conceptual framework. World Bank Policy Research Working Paper 3554. The World Bank, Washington, DC, USA.

Bitzer, V. and Bijman, J., 2015. From innovation to co-innovation? An exploration of African agrifood chains. British Food Journal 117: 2182-2199.

Chavas, J.P., 2004. Risk analysis in theory and practice. Elsevier, San Diego, CA, USA.

Coley, D., Howard, M. and Winter, M., 2009. Local food, food miles and carbon emissions: A comparison of farm shop and mass distribution approaches. Food Policy 34: 150-155.

Darby, M.R. and Karni, E., 1973. Free competition and the optimal amount of fraud. Journal of Law and Economics 16: 67-88.

Donovan, J., Franzel, S., Cunha, M., Gyau, A. and Mithöfer, D., 2015. Guides for value chain development: a comparative review. Journal of Agribusiness in Developing and Emerging Economies 5: 2-23.

Fagerberg, J., 2005. Innovation: a guide to the literature. In: Fagerberg, J., Mowery, D.C. and Nelson, R.R. (eds.) The Oxford handbook of innovation. Oxford University Press, Oxford, UK, pp. 1-27.

FAO/WHO, 2006. Understanding the Codex Alimentarius. Third Edition. FAO, Rome, Italy.

Fritz, M. and Schiefer, G., 2008. Food chain management for sustainable food system development: a European research agenda. Agribusiness 24: 440-452.

Hall, A., Bockett, G. and Taylor, S., 2001. Why research partnerships really matter: innovation theory, institutional arrangements and implications for developing new technology for the poor. World Development 29: 783-797.

Hatanaka, M., Bain, C. and Busch, L., 2005. Third-party certification in the global agrifood system. Food Policy 30: 354-369.

Henson, S. and Reardon, T., 2005. Private agri-food standards: implications for food policy and the agri-food system. Food Policy 30: 241-253.

Henson, S. and Humphrey, J., 2009. The Impacts of Private Food Safety Standards on the Food Chain and on Public Standard-Setting Processes. Paper Prepared for FAO/WHO. Rome: FAO.

Henson, S. and Humphrey, J., 2010. Understanding the complexities of private standards in global agri-food chains as they impact developing countries. Journal of Development Studies 46: 1628-1646.

Humphrey, J. and Schmitz, H., 2002. How does insertion in global value chains affect upgrading in industrial clusters. Regional Studies 36: 1017-1027.

INREF, 2010. Journey into interdisciplinarity. Ten years of INREF experience. Wageningen UR, Wageningen, the Netherlands.

Jaffee, S., Henson, S. and Diaz Rios, L., 2011. Making the grade: smallholder farmers, emerging standards, and development assistance programs in Africa. A research program synthesis. World Bank, Washington, DC, USA. https://openknowledge.worldbank.org/handle/10986/2823.

Lazzarini, S.G., Chaddad, F.R. and Cook, M.L., 2001. Integrating supply chain and network analyses: the study of netchains. Journal on Chain and Network Science 1: 7-22.

Lee, J., Gereffi, G. and Beauvais, J., 2012. Global value chains and agrifood standards: challenges and possibilities for smallholders in developing countries. Proceedings of the National Academy of Sciences of the USA 109: 12326-12331.

Lundvall, B.-A. (ed.), 2010. National systems of innovation. Towards a theory of innovation and interactive learning. Anthem Press, London, UK.

Luning, P.A., Devlieghere, F. and Verhé, R., 2006. Safety in the agri-food chain. Wageningen Academic Publishers, Wageningen, the Netherlands.

Nelson, P., 1970. Information and Consumer Behavior. Journal of Political Economy 78: 311-329.

Oude Ophuis, P.A.M. and Van Trijp, H., 1995. Perceived quality: a market driven and consumer oriented approach. Food Quality and Preference 6: 177-183.

Pietrobelli, C. and Rabellotti, R., 2011. Global value chains meet innovation systems: are there learning opportunities for developing countries? World Development 39: 1261-1269.

Popkin, B.M., 1999. Urbanization, lifestyle changes and the nutrition transition. World Development 27: 1905-1916.

Poulton, C., Kydd, J. and Dorward, A., 2006. Overcoming market constraints on pro-poor agricultural growth in sub-Saharan Africa. Development Policy Review 24: 243-277.

Reardon, T., Barrett, C.B., Berdegué, J.A. and Swinnen, J.F.M., 2009. Agrifood industry transformation and small farmers in developing countries. World Development 37: 1717-1727.

Reeves, C.A. and Bednar, D.A., 1994. Defining quality: alternatives and implications. Academy of Management Review 19: 419-445.

Ruben, R., Van Boekel, M., Van Tilburg, A. and Trienekens, J.H. (eds.) Tropical food chains: Governance regimes for quality management. Wageningen Academic Publishers, Wageningen, the Netherlands.

Saliola, F. and Zanfei, A., 2009. Multinational firms, global value chains and the organization of knowledge transfer. Research Policy 38: 369-381.

Smits, R., 2002. Innovation studies in the 21st century: questions from a user's perspective. Technological Forecasting and Social Change 69: 861-883.

Steenkamp, J.-B.E.M., 1989. Product quality. Van Gorcum, Assen, the Netherlands.

Sumberg, J., 2005. Systems of innovation theory and the changing architecture of agricultural research in Africa. Food Policy 30: 31-41.

Trienekens, J.H., 2011. Agricultural value chains in developing countries: a framework for analysis. International Food and Agribusiness Management Review 14: 51-82.

Van der Grijp, N., Marsden, T. and Cavalcanti, J.S.B., 2005. European retailers as agents of change towards sustainability: the case of fruit production in Brazil. Journal of Integrative Environmental Sciences 2: 445-460.

World Bank, 2007a. World Development Report 2008: agriculture for development. World Bank, Washington, DC, USA.

World Bank, 2007b. Enhancing agricultural innovation: how to go beyond the strengthening of research systems. World Bank, Washington, DC, USA.

2. Linking smallholder farmers to high quality food chains: appraising institutional arrangements

A. Royer[1], J. Bijman[2]* and V. Bitzer[3]

[1]*Université Laval, Département d'économie agroalimentaire et des sciences de la consommation, Pavillon Paul-Comtois, local 4401, 2425 rue de l'Agriculture, Québec G1V 0A6, Canada;* [2]*Wageningen University, Management Studies, P.O. Box 8130, 6700 EW Wageningen, the Netherlands;* [3]*Royal Tropical Institute (KIT), P.O. Box 95001, 1090 HA Amsterdam, the Netherlands; jos.bijman@wur.nl*

Abstract

Although markets for high quality products might represent an interesting outlet for smallholder farmers from developing countries, access to those markets is challenging, as appropriate institutions helping farmers to comply with quality requirements are often missing. To overcome the institutional constraints and to link smallholders to markets, three types of institutional arrangements are often proposed: contract farming, producer organisations and partnerships. While many publications have explored the merits of each of these arrangements, a systematic comparison and evaluation of all three has not been done, particularly from the perspective of the constraints that smallholders face when seeking to improve product quality. In this chapter, we seek to make such evaluatory comparison. To do so, we first identify the most limiting institutional constraints faced by smallholder farmers related to quality improvement. Second, we provide an overview of each arrangement's ability to address these constraints. Third, we determine how combinations of the three arrangements can be used effectively in quality improvement in smallholder value chains.

Keywords: contract farming, producer organisation, partnerships, institutional environment, quality improvement, developing countries, food chains

2.1 Introduction

In both the development economics and the development practice literature, linking smallholder farmers to formal markets is a widely discussed topic. One of the main challenges for smallholder farmers is that formal markets, both domestic and export, have increasingly strict requirements in terms of quality, certification, traceability, minimum quantity, product uniformity, delivery times and food safety (Hatanaka *et al.*, 2005; Reardon and Barrett, 2000). These can be summarised as increased specifications regarding (1) product attributes, (2) process attributes, and (3) transaction attributes (Jaffee *et al.*, 2011). High quality food chains, such as those for fresh fruit and vegetables, fish and fish products, meat, nuts, spices and floriculture, make up a rapidly growing share of international trade in agricultural products from developing country suppliers (World Bank, 2007). National and international formal markets have the potential to increase the welfare of those smallholders that succeed in positioning themselves competitively (Jaffee and Henson, 2005; Maertens and Swinnen, 2009). However, several constraints prevent them from actually meeting the quality requirements (Poulton *et al.*, 2010). In addition to a lack of productive assets, low degree of education and poor infrastructure, smallholder farmers face important institutional constraints that prevent them from producing for and transacting in high quality food chains (Dorward *et al.*, 2005; World Bank, 2007).

In order to benefit from high value formal markets, smallholder farmers need to find solutions to the constraints they experience both in upgrading their production and in accessing input and output markets. The development literature generally distinguishes between three types of institutional arrangements that can help smallholder farmers to overcome those constraints: contract farming (CF), producer organisations (PO), and partnerships.

The first arrangement, CF, is not new for traditional cash crops like cotton, tea and tobacco, but has become more important because of the increasing need for vertical coordination in value chains (Swinnen and Maertens, 2007; Jia and Bijman, 2014). CF can support quality upgrading both by providing farmers with the appropriate inputs and credit, and by providing a guaranteed market for the high value products. The second arrangement concerns collective action in POs. By setting up or joining a bargaining association or cooperative, farmers can reduce the transaction costs resulting from a weak bargaining position and a lack of market information. Although cooperatives and other types of POs have a mixed record of supporting smallholder farmers, they have received renewed attention as suitable organisational solutions in a liberalised economy (Bijman *et al.*, 2016; World Bank, 2007). Partnerships, often referred to as public-private partnerships, represent the third institutional arrangement. Partnerships aim to fill institutional voids by improving the division of labour in global food chains based on the complementarity of different actors. Roles that were traditionally played by one actor can benefit from being shared with, or transferred to, other actors (Kolk *et al.*, 2008; Narrod *et al.*, 2009).

Despite their different core characteristics, all three types of institutional arrangements reflect a degree of alignment between farmers and markets. All three arrangements feature prominently in the recent popularity of value chain approaches among scholars (Gereffi, 1999; Gereffi *et al.*, 2005; Kaplinsky, 2000; Trienekens, 2011), development agencies (Altenburg, 2007; Donovan *et al.*, 2015) and international organisations (e.g. FAO, ILO). Value chain approaches focus on the interactions among the economic actors that together constitute the value chain, on the role of so-called lead firms, and on the opportunities and barriers for smallholders to benefit from participation in modern value chains. In addition, all three arrangements mirror an enhanced market orientation in public policies supporting smallholder farmers. Benefitting from the opportunities offered by domestic and export markets presupposes adequate knowledge of and ability to comply with customers' requirements. Thus, value chains are no longer about selling what smallholder farmers produce, but about what customers demand (Humphrey and Navas-Alemán, 2010). Finally, all three types of institutional arrangements reflect the increased attention to closer coordination between different chain actors. Higher levels of explicit coordination, both vertically between different actors along the chain and horizontally among producers or among various actors outside the chain who support those producers, have become a key condition to access high value markets (World Bank, 2007).

From the great wealth of empirical studies found in the literature, we can conclude that one of the most relevant questions is not whether, but how the different institutional arrangements play a role in linking farmers to markets. Moreover, given the high expectations among practitioners about these arrangements, a literature review of their tangible effects on quality would be most appropriate. This chapter thus aims to make three distinct contributions. Firstly, while there is a growing amount of literature on these institutional arrangements more generally, the three types have not yet been systematically evaluated from the perspective of improving product quality in smallholder value chains. By focusing specially on quality constraints, we seek to analyse in how far the current popularity of the three institutional arrangements in development practice matches the supply chain reality, where alignment between quality produced and quality demanded is a key requirement for market access and competitiveness. Secondly, by comparing the three institutional arrangements, we will provide an overview of their ability to address the institutional constraints to quality improvement faced by smallholder farmers. As we focus on institutional instead of individual constraints, we abstract from the lack of individual assets and capabilities and from individual solutions. Thirdly, rather than understanding each of the institutional arrangements in isolation, the paper seeks to determine how and under what conditions combinations of the three arrangements can be used effectively in quality improvement. These findings will help to identify key areas for development interventions in linking smallholder farmers to high quality markets.

The chapter is structured as follows. Section 2.2 presents our conceptual framework, notably the concepts of institutions and the institutional constraints related to quality improvement and coordination. In Section 2.3, we present the institutional arrangements that we study, namely CF, POs, and partnerships. Each arrangement is assessed from the perspectives of alleviating quality constraints and contributing to (smallholder) development. Section 2.4 identifies the interactions among and possible combinations of the three institutional arrangements. Section 2.5 reflects on the key findings of this literature review and concludes with recommendations for development policy and development interventions.

2.2 Institutional constraints to quality improvement

Although markets for high quality products might represent an interesting outlet for smallholder farmers in developing countries, access to those markets remains challenging, often caused by a lack of appropriate institutions to help farmers comply with quality requirements. Institutions can broadly be understood as the rules of the game in a society; they enable, direct and constrain human behaviour (North, 1990). Institutions create reliability in human interactions and thereby decrease uncertainty in commercial exchange, hence reducing transaction costs[1]. Low cost exchange, in turn, supports specialisation and division of labour, which is one of the foundations of economic development. Throughout the chapter, we use the two terms 'institutional environment' and 'institutional arrangement' (Davis and North, 1971). The institutional environment refers to the macro institutions such as a political order, judicial system, education system, and a certain level of trust in society. Institutional arrangements refer to the modes of organising and coordinating production and exchange activities within a given institutional environment.

In many developing countries, a weak institutional environment entails low information availability, high coordination costs and high risks (Dorward *et al.*, 2003). In addition, many farmers, particularly those in remote areas, face inadequate transport and storage facilities, and poor telecommunication networks. Poor physical infrastructure significantly raises transaction costs and negatively affects the production and marketing options of farmers (Barrett, 2008; Fafchamps and Hill, 2005). This is particularly true for farmers seeking to produce high quality products as these are often perishable products. Poor infrastructure thus leads to problems of quality deterioration and high food losses (World Bank, 2007). Unfortunately, in many developing countries, especially in Africa, government investment in physical infrastructure and other key public goods in rural areas has declined over the past decades (Jayne *et al.*, 2010).

Next to a weak institutional environment, smallholder farmers face constraints that relate to their access to input and output markets, to credit and information, and to technical assistance and innovation options. Such access is severely constrained, as markets do not function well, the quality of extension is low, and necessary (market) information is lacking. Again, farmers seeking to produce for high value markets are more likely to be constrained by the lack of enabling institutions. We use the term 'institutional constraint' to emphasise that solutions cannot be expected from individual farmers, but need some form of collective action. Figure 2.1 shows the various institutional constraints experienced by smallholder farmers. On the input side, these constraints relate to the lack of inputs, financial services and extension services. On the output side, lack of market information, quality control and inspection,

---

[1] Transaction costs are considered as the costs of contracting, of performing an economic exchange. They consist of *ex ante* and *ex post* costs. *Ex ante* costs are basically search, bargaining and contract writing costs. *Ex post* costs are enforcement, monitoring, maladaptation and contract breach costs.

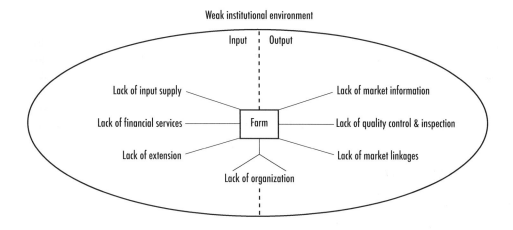

Figure 2.1. Institutional constraints faced by smallholder farmers.

and linkages to output markets constitute the main institutional constraints. The lack of horizontal organisation (i.e. among producers) concerns both input and output markets. We will now discuss each of these seven institutional constraints in more detail.

The first constraint is the lack of financial services, including credit, savings and insurance. Formal credit institutions are usually out of reach for smallholders due to the lack of collateral and the informal nature of most trade arrangements. This makes these producers high risk borrowers, leading to excessively high interest rates, as lenders must substitute costly monitoring for the lack of collateral (World Bank, 2007). In addition, transaction and monitoring costs are particularly high as farmers are geographically dispersed, transactions are small and infrequent, and risks of moral hazard and covariance (e.g. related to weather conditions) are considerable (Kirsten *et al.*, 2009; Poulton *et al.*, 2010).

The consequence of difficult access to finance is a low ability to make the necessary investments for switching to other crops, other varieties or other production methods. Non-traditional high quality crops require significant investments in inputs, improved production and processing technology, and technical expertise (e.g. Jaffee and Masakure, 2005; Key and Runsten, 1999). In a study on fresh vegetable producers in Kenya, Muriithi *et al.* (2011) found that the lack of capital to cover the high costs of certification with GlobalGAP and the high costs of the recommended inputs acted as a major constraint for compliance with customers' demands. This confirms earlier findings that especially for 'risky' products, such as perishable fresh fruit and vegetables, smallholders face great difficulties in making the needed investments by themselves to link up with formal markets (e.g. Berdegué *et al.*, 2005).

A second institutional constraint is the lack of inputs, such as fertilisers, high quality planting material, pesticides, equipment, and growth regulators. High quality input material is a crucial prerequisite for farmers to achieve high product quality. However, there are fundamental problems with supply and demand. Liberalisation of state controlled input supply systems and recent cutbacks in public seed distribution systems have reduced the quality and quantity of inputs offered by state agencies, whilst the private sector has not always stepped in to fill the gap to cater for the needs of smallholder farmers (World Bank, 2007). At the same time, demand for inputs by smallholders is low and uncertain due to farmers' lack of cash liquidity and the small scale of transactions. Furthermore, high input prices combined with uncertain output prices undermine the benefits of high quality inputs (Kydd and Dorward, 2001: 473). Such constraints hinder smallholder farmers from enhancing productivity and improving the quality of their product to meet market requirements.

A third constraint is the lack of (appropriate) extension and technical assistance. Agro-food chains have evolved rapidly in recent years, and demands on suppliers have risen substantially. However, smallholder farmers often lack the knowledge about optimal production, harvesting, processing and storage techniques, especially with regard to high quality crops, as these are often new to farmers. Agricultural extension services have the task to translate knowledge generated at research institutes to the practice of farmers. Accordingly, smallholder farmers with access to extension services have been found to be more likely to adopt new technology and invest in new market trends, such as certification (Asfaw *et al.*, 2010; Muriithi *et al.*, 2011). However, public extension services in developing countries only reach a minority of smallholder farmers and are often not tailored to their needs, strategies and resource constraints (Poulton *et al.*, 2010). In addition, extension agents often lack relevant crop-specific knowledge and have weak incentives to make real impact (Spielman *et al.*, 2010; World Bank, 2007). Thus, public services do not fill the need for technical assistance in a way that is adequate to improve smallholders' production and processing practices to enter high quality product chains (Berdegué *et al.*, 2005).

A fourth constraint is the lack of horizontal organisation among smallholders, due to lack of trust, lack of organising skills and relatively high governance costs of collective organisations. On an individual basis, smallholder access to high quality markets is constrained by small and infrequent transactions, aggravated by generally low bargaining power (Dorward *et al.*, 2003). Hence, some kind of collective action is often needed to benefit from economies of scale and scope and to reduce transaction costs (e.g. Markelova *et al.*, 2009). POs could reduce the transaction costs associated with accessing inputs, information, technology, and credit, and they can assist with processing and marketing activities. This enables farmers to compete with larger producers of high quality products, and improves their bargaining power vis-à-vis buyers. Section 2.3.2 will further elaborate on the pros and cons of POs.

A fifth constraint is a lack of market information. Transacting in high quality value chains requires having information on customers' demands, prices, grade specifications, storage and transport facilities, etc. However, in many developing countries, market information systems are absent or performing poorly due to inadequate financing and unreliable and inaccurate data (Gabre-Madhin, 2009). Existing services are usually limited to low-value commodities, and are ill-equipped to provide information on high quality products for domestic or export markets (Shepherd, 2007). Lacking market information may prevent smallholders from making investments in specific production techniques to comply with customer demands. Okello and Swinton (2007) report that the lack of knowledge of farmers about their own product quality also leads to problems of price holdup, as buyers are able to reject products based on supposed quality problems, especially during periods of oversupply.

A sixth institutional constraint for farmers producing for high quality markets concerns the lack of quality control and inspection services. In order to participate in these markets, farmers must be able to demonstrate compliance with specific production and harvesting protocols. Access to certification, auditing, quality control and laboratory services might be critical to verify assured compliance with buyers' requirements (Jaffee *et al.*, 2011). However, auditing and certification services are expensive as they have strong economies of scale (Raynolds, 2004). Many countries lack the technical facilities and service providers necessary for quality control. Thus, countries with established quality support services have been found to have a comparative advantage when it comes to complying with changing consumer and retail preferences. Henson *et al.* (2011) found that firms in countries with existing quality control services are more likely to have achieved GlobalGAP certification for horticultural exports as compared to countries that have newly established control services. Similarly, Jaffee *et al.* (2011) state that one of the key problems with efforts to scale up projects promoting the use of private standards in developing countries is the inadequate availability of standards-related services.

The final institutional constraint relates to the lack of market linkages, which is often caused by the remoteness and dispersion of rural production and by low production volumes of individual farmers. Market linkages may also be hindered by monopolistic practices, corruption, and low market trust. Moreover, the quality of smallholders' products is often uneven and uncertain, making private companies reluctant to procure from smallholder producers. Studies suggest that supermarkets' procurement practices are often biased in favour of large and medium scaled suppliers given the lower costs and risks of doing business with them (Berdegué *et al.*, 2005; Weatherspoon and Reardon, 2003).

In conclusion, each of the seven institutional constraints described above has serious implications for the opportunities that smallholders have in benefitting from quality upgrading. Table 2.1 summarises the main institutional constraints and their effect on quality improvement.

It needs to be emphasised that these institutional constraints are interrelated and even interdependent. For instance, credit is often needed to buy inputs, while technical

Table 2.1. Main institutional constraints for quality improvement.

Institutional constraint	Effect on quality improvement (QI)
Lack of financial services	QI requires additional financial investments in inputs, cultivation practices and crop handling
Lack of inputs	QI requires additional and often specialised inputs, e.g. seeds, fertilisers and pesticides
Lack of extension services	QI requires specialised technical advice and training
Lack of organisation	QI requires the reduction of vulnerability of smallholders farmer to transaction risks
Lack of market information	QI requires the availability of information on market demands and on customer requirements
Lack of quality control and inspection services	QI requires effective and affordable quality control systems
Lack of market linkages	QI requires the availability of market linkages to create outlets for smallholders' products

assistance is necessary for effective and efficient use of those inputs. Access to high quality markets requires on-farm investments that only pay back if farmers are able to demonstrate compliance with quality standards through certification. However, obtaining certification may be difficult without support from extension services or other providers of technical assistance, and without access to credit. Hence, the solution to these institutional constraints calls for complementary coordination of market activities (Dorward *et al.*, 2003; Kydd and Dorward, 2004). Input- and output-related services to smallholder producers need to be coordinated 'so that individual investors are assured that their investments will not fail as a result of other investors either failing to make complementary investments or behaving opportunistically' (Poulton *et al.*, 2006: 266). Only when several institutional constraints are solved simultaneously will smallholder farmers be able to improve quality and successfully participate in high quality agrifood chains.

2.3 Evaluating institutional arrangements from a quality improvement perspective

The complementarity of the institutional constraints calls for concerted action to ensure that coordination problems are solved. As Kydd and Dorward (2004) argue, the necessary action will not be achieved by market mechanisms alone, especially not in rural areas with thin markets. Therefore, different institutional arrangements have emerged as solutions. Among these arrangements, CF, POs and partnerships have been used substantially in development practice and have received considerable attention from academics and policy makers. Each of these arrangements may reduce institutional constraints by establishing market linkages (e.g. linking buyers and producers, assisting in negotiation, providing information on quality requirements), setting up training and technical assistance, supporting the development of POs, providing access to input markets including credit, and facilitating quality assurance and certification. However, a debate among both academics and practitioners persists on the question what type of institutional arrangement is most promising for integrating smallholders in high quality value chains and thereby providing better income opportunities for smallholder farmers. More specifically, the debate centres on the question what institutional arrangement is most appropriate under what technical, social and economic conditions. Based on a review of the literature, we will discuss, in the following subsections, each of the three institutional arrangements, their key characteristics, and their impact on quality improvement.

2.3.1 Contract farming

CF arrangements or outgrower schemes[2] form one type of institutional arrangement that can be used to improve the quality of farm products. More specifically, CF can be a solution to the various institutional constraints for quality upgrading as listed above. CF has been defined as an agreement between one or more farmer(s) and a contractor for the production and supply of agricultural products under forward agreements, frequently at predetermined prices (Eaton and Shepherd, 2001). The contractor is usually a trading company or a processing company. By contracting with (smallholder) farmers, these buyers make sure they will receive the right quantity and quality for the processing plant or for fulfilling downstream contractual agreements with retailers.

CF arrangements have three distinct functions (Hueth *et al.*, 1999; Sykuta and Cook, 2001; Wolf *et al.*, 2001). First, they serve as a coordination device, allowing each chain actor to make decisions (e.g. on resource allocation) that are aligned with the decisions of the other actors. Coordination is meant to ensure that products of the right quantity and quality are produced and delivered at the right time and place. Second, contracts are used to clarify and agree upon the incentives and penalties that motivate performance. Without proper incentives for each contract partner, no transaction will take place. Particularly when the contractor demands specific investments from the farmer, for instance in the case of high quality, the contract clarifies what compensation the farmer will obtain for this investment. Third, the contract arranges the allocation of risk. As producer and buyer usually have different risk preferences, the contract can shift risk from the risk-averse to the risk-neutral partner. For example, farmers can mitigate the risk of income loss due to poor yield by signing an agreement with a contractor that specifies a portion of compensation independent of the yield realised.

CF is often presented as an institutional arrangement for organising vertical coordination in the value chain, notably between producers and their buyers (Jia and Bijman, 2014; Swinnen and Maertens, 2007). Vertical coordination means that

---

[2] In most of the literature, 'outgrower schemes' and 'contract farming arrangement' are used interchangeably; we will also use these terms as synonyms.

the activities of producers and buyers are closely aligned. As food value chains are characterised by sequential transactions, vertical coordination implies that the transaction upstream (e.g. between producer and processor) is aligned with the transaction downstream (e.g. between processor and distributor). In food chains, upstream and downstream transactions have become increasingly interdependent because processors and distributors use consumer brands to strengthen their competitiveness. To protect a brand from devaluation, processors and distributors seek to control those transactions upstream in the value chain that can affect the value of the brand. Thus, CF is a tool for vertical coordination, not only between farmers and their primary customers, but for the whole food chain.

More vertical coordination between seller and buyers is particularly needed for high-value products, highly perishable products and for technically difficult products (Goldsmith, 1985; Minot, 1986). First, for products with high value, variation in quality has important economic implications. Customers of these products are often willing to pay a premium for a specific product, variety or attribute, as well as for consistent quality. This premium covers the additional cost of producing and the cost of the CF arrangement. Often, farm-level investments in human and physical capital, or specialised inputs are needed to obtain the desired quality level. CF can then provide farmers the incentives and the means to make these specific investments. Second, highly perishable products require intense coordination between production, harvesting and processing/marketing. In addition, perishable products entail high transaction risks. Without some contractually guaranteed outlet, the farmer is not likely to produce such perishable products. Third, farmers are not likely to enter into the production of technically difficult crops when they do not have the technical skills, the inputs and the credit needed. As part of a CF arrangement, buyers can provide technical assistance, specialised inputs and credit.

Thus, the need for vertical coordination in the producer/contractor relationship is highest when products are highly perishable, when the production requires specific investments, and when the production involves special skills and inputs. These conditions are most likely to be present when the buyers are large-scale processing companies, exporting firms, and suppliers of modern supermarkets. In general, foreign markets, particularly those in developed countries, demand products to comply with high quality and food safety standards, thus providing an incentive for traders to closely monitor production processes. CF is the arrangement commonly used to be able to guarantee the quality of the products.

In sum, customer demand for higher quality and for quality assurance leads to more vertical coordination in value chains. CF has become a popular arrangement to organise the strict coordination needed. We now move to answering the question how CF arrangements can solve the institutional constraints that are often inhibiting quality improvements by smallholder farmers:
- *Financial services.* Contracts can provide smallholders with access to credit. Being more costly to produce, high value crops often require credit to purchase additional

and specialised inputs, to hire additional labour for planting and harvesting, and to buy or rent specialised equipment. Since contracting firms generally have more financial capacities than smallholders, they often provide their contractees with the necessary credit. Reardon *et al.* (2003) note that even if the contracting firm does not provide credit directly, banks will generally accept the contractual commitment as collateral for granting loans.

Key and Runsten (1999) argue that contracting firms are well suited to act as lenders to growers, because they have a superior ability to monitor and enforce credit contracts and have lower default costs than banks. Contractors can extract a grower's debt directly from the crop revenue before the grower receives his payment. Because alternative markets for high value products are often thin, growers have no choice as to sell to the contractor. At the same time, the contractor can be assured that the loan will be spent on production because loans are usually distributed in kind or in the form of vouchers and the contractor often monitors the use of inputs.

- *Input supply.* One of the key elements of CF in developing countries is the interlinkage of inputs and output markets (Dorward *et al.*, 1998). CF can provide smallholder farmers with inputs that might otherwise be too costly or unavailable (Key and Runsten, 1999). High-value products cannot be produced without using proper seeds and planting material, without applying the proper inputs (fertilisers and crop protection chemicals) at the right time, and without deploying the proper planting, spraying and harvesting machinery. The contracting firm, such as a processor, can make these inputs available directly or can contract with one or more input supplying firms. Boselie *et al.* (2003) found that even supermarkets may provide their supplying producers with inputs in order to obtain the right quality.

Provision of inputs by the contractor reduces the transaction costs related to grower uncertainty about the availability and quality of inputs. Failures in input markets are circumvented by such direct provision, and the economies of scale allowed by the large volume purchase of inputs by the contractor can be passed on to farmers, which may lead to a reduction in the price the contractor has to pay for the farm products.

- *Extension services.* CF schemes commonly include provisions on technical assistance and advisory services. As high-value crops generally need more and newer skills than the grower has at hand, technical advice is crucial for inducing smallholder farmers to grow these crops. Most contracts described in Glover and Kusterer (1990) included visits by firm extension officers to either individual farmers or farm groups several times during the first year of the contract but often less in later years. Boselie *et al.* (2003) reported that some contracting firms apply all pesticides and hire extension officers that give advice on production practices. Often, contractors communicate product and technology information to growers by firm-employed extension agents. These agents not only give technical advice to the growers, but can also monitor their behaviour (Hueth *et al.*, 1999).

By providing technical assistance to farmers, contractors can obtain more uniform products, which is important in their quality-oriented marketing strategies.

Without such assistance, farmers may not be willing or able to venture into innovative crop and livestock enterprises as these involve higher risks. At the same time, this technical assistance can enhance the management skills of the farmer and there may be spill-over effects to non-contracted crop and livestock activities.

- *Organisation.* CF as such does not solve the constraint of lacking organisation. However, contractors may require their suppliers to form groups, so they can economise on transaction costs in dealing with farmers. Such groups can also be existing collective action groups. Still, most contractors are not eager to support the formation of groups among their suppliers. The benefits of dealing with a group instead of with many individuals may be offset by the disadvantage of the group becoming a bargaining organisation, demanding higher prices or better delivery conditions for the growers.

  Several studies claim that farm groups, such as formal or informal POs, may support the efficiency and equity of CF (Glover, 1984; Coulter *et al.*, 1999; Key and Runsten, 1999). POs can improve the power balance between producers and contractors, thereby strengthening the incentives for both parties to continue bilateral contracting. In addition, POs can reduce the transaction costs in the contract arrangement, as the contractor does not have to deal with numerous smallholder farmers but with only one organisation of smallholders. Finally, POs may support CF by channelling and supporting (e.g. by providing legitimacy to) the technical assistance needed to help producers increase product quality and uniformity.

- *Market information.* Information is a key element for achieving higher quality. In order to meet market requirements, farmers must be aware of the chemicals permitted, residue levels allowed, desired product characteristics (texture, shape, flavour, colour, etc.) and ideal timing of the harvest (Key and Runsten, 1999). Markets can transfer information about supply and demand via prices, but prices usually cannot rapidly communicate complex and changing quality demands. Contracts generally include provisions on the transmission of information about final market demands from contractor to farmers.

- *Quality control and inspection services.* Some form of quality control is always part of the contracted transaction. Depending on the provisions in the contract, this control is more or less strict. Strict quality control may disadvantage growers, as the products supplied may vary in quality and the grower still want to sell all harvested products (Jaffee *et al.*, 2011). However, transparent quality control linked to quality-dependent pricing may induce farmers to produce better quality, and is particularly in the interest of those farmers that are able to produce the highest quality. Ideally, quality control and inspection are carried out by independent third parties, but in reality this is often not the case due to the high cost of third party involvement.

- *Market linkages.* CF arrangements are particularly about market linkages, at they provide growers with guaranteed market access and it provides contractors with guaranteed supply. If market linkages are weak because of underdeveloped physical infrastructure, CF cannot solve the constraint. Often, the market linkages established by CF are conditional on producers being able to comply with various quality requirements. Thus, CF cannot solve the lack of market linkages for all smallholder farmers, particularly not for those in remote areas.

Several studies have empirically investigated the direct impact of CF on the improvement of quality. In a study on the contracting practices of smallholder vegetable farmers in Madagascar, Minten *et al.* (2009) observe that given the right incentives and contracting system, small farmers can participate successfully in emerging high-value markets. Farmers fulfilled the complex quality requirements of an exporter to foreign supermarkets through the use of micro-contracts combined with extensive farm assistance and supervision programmes. The exporter employs 300 extension agents to supervise the production of almost 10,000 suppliers. The authors conclude that farmers participating in these contracts have higher welfare, more income stability and shorter lean periods than before they started contracts.

Another success story was reported from Mexico by Key and Runsten (1999). A processing firm succeeded in achieving the required quality with smallholders because it was able to offer contracts that provided for credit, specialised inputs and extension assistance at low transaction costs. Boselie *et al.* (2003) examined five cases of supply chains for fresh horticultural products sold to African, Asian and European supermarkets. These cases show that small producers were generally able to comply with supermarket requirements by using contracts that supplemented for weak public institutions. Henson and Reardon (2005) affirm that small-scale producers are able to achieve supermarket quality levels of at least equal to those of large-scale producers through contracts on the condition that well-coordinated systems of control of smallholders are in place. These systems need to include measures that facilitate compliance, incentives to reward good performance, and penalties to punish non-compliance.

Although improving smallholders' access to quality markets, CF should not be considered a cure-all for smallholders. Some scholars mention that by providing credit, CF makes farmers dependent on their contracting partners that could keep them in perpetual indebtedness (Little and Watts, 1994; Singh, 2002). It has been also been argued that contracts give rise to uneven bargaining power between partners (Glover, 1984; Little and Watts, 1994). Also, contracts may not be renewed for the next season, which might lead to the loss of value of farmer investments specifically made for the contracted transaction.

2.3.2 Producer organisations

A PO is a private collective action organisation established by agricultural producers (often with third party support) in order to support their economic well-being (Bijman *et al.*, 2016). Broadly, POs may fulfil three functions for their members: provide services when markets fail, provide club goods or local public goods when states fail, and provide a voice in political affairs (Rondot and Collion, 2001). Taking the public good and voice functions together, Thorp *et al.* (2005) have made the distinction between claims groups and efficiency groups. Claims groups typically operate vis-à-vis public authorities as they seek to get favourable conditions from policy makers and administrators, such as subsidies, enabling legislation or other public benefits for group members. Claims groups are lobby organisations and advocacy organisations. Efficiency groups seek to increase the efficiency of the production and marketing processes of farmers, for instance by reducing transaction costs and improving bargaining power. In this chapter, as in the rest of the book, we focus on POs that have primarily an economic function, by providing inputs to the farmer-members and/or selling the products of the farmer-members. While efficiency groups often also engage in lobbying, it is not their primary function.

The difficulties of small farmers in accessing high-value markets has prompted a renewed interest for the role of POs in helping producers to improve the quality of their products (World Bank, 2007). Many authors have mentioned the potential role of collective action organisations in helping farmers to face challenges imposed by high-demanding markets (Brester and Penn, 1999; Hellin et al., 2009; Markelova et al., 2009). In order to understand how exactly POs solve the institutional constraints identified in the previous section, we now identify the functions that POs can and often do perform to help alleviate them.

A strong justification for the existence of POs is their capacity to facilitate access to inputs. First, POs facilitate input access for farmers through bulk purchase, which lowers prices, or affiliation with larger group members (Kaganzi et al., 2009). Second, the resources of POs are especially beneficial to smallholder when high investment costs are required to enter specific markets such as irrigation, storage, or cold chain, often necessary inputs to the production of high-quality products. Third, observing that physical availability of inputs is often an important constraint to access, with thin and unreliable rural distribution networks in most African countries, Kindness and Gordon (2001) claim that POs can act as a vehicle for input distribution. Dorward et al. (2004) emphasise the effectiveness of POs in coordinating the provision of various services to smallholder producers.

POs can facilitate access to finance in various ways. Farmers may become members of a dedicated Savings and Credit Cooperative, or their own supply or marketing cooperative provides them with credit. Kaganzi et al. (2009) reported on a case study of potato farmers in Uganda who pooled their financial resources from personal savings and loans to establish a savings and credit cooperative so as to save and access loans to invest in production of high quality potatoes.

POs can hire technical experts and develop their own extension services. Moreover, when supplied adequately, these services can be better tailored to members' needs than public extension services. For instance, cooperatives often provide technical assistance together with the inputs (such as seeds and pesticides) that they provide to their members (Markelova et al., 2009). However, the literature on the impact of POs on access to extension services is mixed as to the outcomes. The infant stage of services provided by POs may not be adequate for farmers that adopt quality-enhancing techniques. In a study on Ethiopian dairy marketing cooperatives, Francesconi (2009) found that cooperative membership had a negative impact on milk quality compared to non-members because members adopted productive crossbred cows that were more sensitive to mastitis than the local breeds. Crossbred cows require more sophisticated feeding and husbandry techniques than the indigenous cattle and require support that the cooperative services had difficulty to provide.

Organise individual producers is by definition the objective of POs. For quality improvement, organising farmers in POs can reduce the risks that individual farmers face when engaging in high-value products (Shiferaw *et al.*, 2006). Verhaegen and Van Huylenbroeck (2002) argue that collective action is an important institutional arrangement for smallholders since it reduces the risks of individual farmers to invest in the costly assets that are needed to improve quality and that are often specific to a particular value chain. Raynaud and Sauvée (2000) mention that only cooperation among producers can sufficiently reduce costs of the necessary investments in specific skills and assets to improve quality. POs can also create economies of scale necessary for smallholders to be competitive in accessing high-quality markets (Biénabe and Sautier, 2005). Collective marketing allows farmers to spread the higher costs of marketing high-value products.

POs could reduce transaction costs by collecting market information, processing it and sharing it among its members (Markelova *et al.*, 2009). Although most POs do collect and process information about the demands and requirements of their customers, there is little evidence that POs are a central vector for the provision of more general market information. A reliable market information system requires substantially more resources than an individual PO has available. Market information is a typical public good, and individual POs are not likely to invest in setting up a system that other (non-member) farmers will also benefit from.

POs often play a role in quality control and inspection. By checking the quality of the products that members deliver and sorting them in various quality classes, POs reduce the information asymmetry between seller and buyer and thus increase the efficiency of the sales process. The literature provides many examples of cooperatives involved in quality control (e.g. Francesconi *et al.*, 2010; Moustier *et al.*, 2010). Cechin *et al.* (2013) even found that cooperatives in Brazil maintain higher quality standards than their non-cooperative competitors. However, quality control in POs in developing countries is a challenging task. Often, POs do not have the resources and skills to do quality control properly, or members delivering below-standard quality are not sanctioned due to internal politics. Finally, many cooperatives, for instance in dairy, work with delivery groups, and all members of a group receive the same price for the average quality of the group. This group system allows for opportunistic behaviour of members.

Facilitating market access is perhaps the most important function of POs, suggested by this often-used definition of a PO: 'A producer organisation is a rural business that is generally owned and controlled by small-scale producers and engages in collective marketing activities' (Penrose-Buckley, 2007). Holloway et al. (2000) found that POs of smallholders in India were able to increase smallholder access to higher value markets by reducing transaction costs. Several authors have found that the rise of supermarkets in developing countries entails an incentive for smallholders to set up joint marketing organisations (Narrod et al., 2009; Vorley et al., 2007; Weatherspoon and Reardon, 2003). The quality standards that supermarkets apply not only induce farmers to raise product quality, but also to supply large quantities of uniform product. POs, by gathering products of a large number of dispersed smallholders, can respond to these customers' needs. POs may also shorten the supply chain by not only selling the product, but also transporting, sorting and packaging the product. A shorter chain means a lower probability of quality deterioration, but also a larger share of the total chain added value for the producer.

The use of POs for producing and marketing agricultural products does not automatically lead to enhanced quality. The decentralised decision-making process within POs can lead to free-riding in quality upgrading as recognised by numerous scholars (e.g. Cook, 1995; Winfree and McCluskey, 2005). POs are also subject to internal organisational transaction costs, notably the costs of communicating the benefits of collective action and coordinating smallholders along these precepts. Besides these organisational problems, POs in developing countries face other challenges, such as poor management by their leaders, low financial means and a rigid institutional environment. Regarding the low financial means, the study of Roy and Thorat (2008) on an Indian cooperative assistance organisation shows that POs, similar to individual producers, often lack human capital and financial means to assist smallholders. They often rely on support from the state or from NGOs. This support, in turn, can create a dependence of POs on outside assistance that can lead to organisational failure when the support is withdrawn.

2.3.3 Partnerships

The key feature of partnerships is the complementarity of actors. Since no single organisation has the capacity (in terms of necessary financial, technical and human resources) to address all institutional constraints unilaterally, partnerships create a structure so that different actors compensate for different institutional constraints. This also suggests that the more numerous the institutional constraints, the higher the need for partners from multiple backgrounds (Rivera-Santos et al., 2002). In addition, partnerships facilitate the spreading of risks among different actors. For instance, in a situation where exporting companies are unwilling to procure from a large number of smallholders, other actors may be able to safeguard companies' investments (Narrod et al., 2009). Especially NGOs and donor agencies are able to fulfil important activities and compensate for institutional gaps (Rivera-Santos et al., 2002). Instead of becoming chain actors, they rather act as facilitators providing advice and mitigating risk.

Partnerships have emerged since the late 1990s as important mechanisms for addressing rural development challenges, promoting capacity building and market access for smallholder farmers, and developing sustainable economies (Glasbergen, 2007; Kolk *et al.*, 2008). Particularly traditional export chains, including coffee, cocoa and palm oil, and more recently high value chains, such as fresh fruit and vegetables, have experienced a considerable proliferation of partnerships (Fuchs *et al.*, 2011; Schouten and Glasbergen, 2011). These can be defined as collaborative, institutionalised arrangements between actors from two or more sectors of society – market, state and civil society – which aim at the provision and/or protection of collective goods (Schäferhoff *et al.*, 2009). Partnerships can take a variety of forms, including public-private partnerships, business-NGO alliances, or multi-stakeholder initiatives.

The development literature examines partnerships in the context of a paradigm shift in the political economy of international development (Van Tulder and Fortanier, 2009). Firstly, partnerships are approached as institutional arrangements to overcome the inability of individual actors to solve the challenges faced by smallholder farmers (Kolk *et al.*, 2008). The solution to single-actor failures is seen to lie with multi-actor collaboration where different resources and knowledge can be coordinated 'to increase the effectiveness of each partner's effort' (Van Tulder and Fortanier, 2009: 229). Secondly, businesses are increasingly called upon to make a positive contribution to international development and poverty alleviation given their key position and resources in global chains. This is most pronounced in global food chains, where the dominant market position of multinational companies makes them 'obvious candidates for collaboration' (Springer-Heinze, 2007: 10) to implement chain-wide innovations and connect smallholder farmers to global markets (Altenburg, 2007).

Hence, strategies to improve smallholder market access increasingly involve partnerships between public and private stakeholders (Boselie *et al.*, 2003). While the donor discourse tends to emphasise the potential of partnerships to improve the position of farmers in global food chains through supporting the application of quality standards and certification (Springer-Heinze, 2007; World Bank, 2007), there are only few studies on the impact of partnerships as institutional arrangements for quality improvement. Those studies reveal considerable differences in the extent to which partnerships address the institutional constraints.

For instance, with regard to improving access to inputs, no meaningful conclusion can be drawn on the performance of partnerships. Although there are cases where partnerships have ensured that farmers receive better access to inputs, for instance seeds or seedlings (Perez-Aleman and Sandilands, 2008; Van Wijk and Kwakkenbos, 2012), most of the literature does not mention partnerships as important providers of inputs.

Similarly, the track record of partnerships to provide access to finance is mixed. Three ways of arranging financial services by partnerships can be distinguished (Bitzer *et al.*, 2011). Partnerships can provide direct credit to farmers, mostly through donor sources, they can establish interlocking credit arrangements between farmers and (micro) finance institutes, or they can promote chain internal credit provision by commercial buyers. The majority of the literature confirms that partnerships provide direct financial support to farmers for the duration of the partnership. This is often done in order to cover the certification fees (Boselie *et al.*, 2003; Narrod *et al.*, 2009). Regarding any longer-term provision of finance, the evidence appears contradictory. Whilst credit facilitation through arrangements with external actors, such as banks or social lending institutes, is rarely documented in the literature, some studies find that partnerships frequently ensure the provision of finance through participating buyers (Ferroni and Castle, 2011; Van Wijk and Kwakkenbos, 2012). Other authors conclude that long-term access to credit is one of the aspects that partnerships seem to be paying little attention to (Bitzer *et al.*, 2008).

The literature clearly indicates that the provision of extension services, i.e. training and technical assistance, constitutes the key focus of partnerships. In most cases, this serves the purpose of supporting smallholder farmers to comply with the product and process standards required to gain access to high quality agrifood chains (Kersting and Wollni, 2011; Okello and Swinton, 2007). Partnerships also help farmers to apply good agricultural practices, enhance production efficiency, raise product quality, and overcome adoption constraints to new technology (Bitzer *et al.*, 2008; Van Wijk and Kwakkenbos, 2012). Several studies report that partnerships can train service providers to serve smallholders (Narrod *et al.*, 2009; Perez-Aleman and Sandilands, 2008; Van Wijk and Kwakkenbos, 2012). Hence, partnerships present a new source of technological change in agrifood chains, thereby fulfilling a task that is generally thought to be the responsibility of governments and public extension services (Bitzer *et al.*, 2011). Especially NGOs assume an active role within partnerships as the main providers of technical assistance (Boselie *et al.*, 2003; Jaffee *et al.*, 2011).

By providing technical assistance and focusing on meeting stringent quality requirements and certification demands, partnerships create strong incentives for the organisation of individual producers into formal groups. Case studies show that partnerships promote group structures to serve as basic training and certification units (Bitzer *et al.*, 2011) and to gain economies of scale in purchasing inputs and in marketing (Narrod *et al.*, 2009; Perez-Aleman and Sandilands, 2008). Thus, partnerships reflect the increased importance of POs in development cooperation as instruments to promote rural development and facilitate smallholder inclusion into markets.

Closely related to technical assistance to smallholder producers is the provision of market information by partnerships. Information focuses on market requirements, such as certification, quality demands, volumes and delivery schedules (Narrod *et al.*, 2009). However, it remains unclear to what extent partnerships also transmit information on prices. Whilst some studies suggest that partnerships may well be able to distribute price-related information (e.g. Thiele *et al.*, 2011), other research does not find partnerships to be distributing information on prices due to the participation of private sector actors reluctant to share this kind of information with producers (Haantuba and de Graaf, 2008).

Given the importance of quality control and inspection services, several partnerships are active to train farm advisors, consultants and group leaders in export requirements and standards, so that they can offer advice and training to smallholder producers (Kersting and Wollni, 2011; Narrod *et al.*, 2009). Other partnerships have created local business services for quality assurance (Perez-Aleman and Sandilands, 2008). Most well-known in this regard is the case of AfriCert in Kenya, which was set up jointly by donors and NGOs as Africa's first indigenous certification company. Finally, there are several examples of partnerships that have developed collective private standards, including certification procedures and quality control systems (Bitzer and Glasbergen, 2010; Schouten and Glasbergen, 2011).

In addition to channeling support and information to producers, partnerships take on a transactional function to increase market access and create new market opportunities for smallholder farmers. The idea is to link farmers to participating companies, often through NGO facilitation and coordination, based on the premise that these companies are interested in developing a stable supply base (Perez-Aleman and Sandilands, 2008). Due to the active participation of large MNCs in partnerships compared to businesses from developing countries partnerships are generally aimed at high value export markets, particularly at certified markets under sustainability schemes or specialty markets. There are, however, also cases of partnerships establishing direct linkages to retailers in developing countries in order to supply the domestic market (Boselie *et al.*, 2003; Haantuba and De Graaf, 2008). Whilst most studies mention the increased market access as one of the key outcomes of partnerships, experiences reveal several hindrances to the establishment of long-term business relations. Firstly, purchase agreements facilitated by partnerships are mostly on an annual basis only (Bitzer *et al.*, 2011). Secondly, market linkages can only be maintained when sufficient quantities and quality are guaranteed, otherwise they collapse. Finally, a lack of trust between public and private parties or between producers and buyers may complicate the development of long-term purchase commitments (Chitundu *et al.*, 2009; Haantuba and De Graaf, 2008).

2.3.4 Comparative analysis

Based on the comprehensive literature review presented above, we have compiled an overview of the evidence of the three institutional arrangements' competences in addressing the institutional constraints faced by smallholder farmers related to quality improvement (Table 2.2). Evidence is here interpreted in terms of whether or not we found substantive evidence in the empirical literature.

Table 2.2 clearly shows that all three arrangements have different competences in addressing institutional constraints. The literature shows strong evidence that CF

Table 2.2. Empirical evidence that the institutional arrangement addresses the institutional constraints related to quality improvement.[1]

Institutional constraints	Contract farming	Producer organisations	Partnerships
Lack of inputs	++	++	−
Lack of financial services	++	+	+/−
Lack of extension services	+	+/−	++
Lack of organisation	−	++	+
Lack of market information	−	−	+/−
Lack of quality control and inspection	++	+	+
Lack of access to market linkages	++	++	+

[1] ++ strong evidence found in the literature; + some evidence found in the literature; +/− inconclusive or contradictory evidence found in the literature; − little or no evidence found in the literature.

addresses the lack of inputs and lack of credit. Also extension (or technical assistance) is often provided as an element of the contractual arrangements. Finally, the contract provides the farmer with a market, and buyers are keen on inspecting the quality of the products delivered by the farmers (although this inspection is not always done in a transparent way). The literature on CF shows contradictory evidence related to the lack of organisation, some results demonstrating a positive effect of contracts on smallholders' organisation while others are more critical. In sum, the key benefits of a CF arrangement is that it provides smallholders with access to inputs, access to credit and access to a market.

As for POs, the literature clearly indicates their role in addressing the lack of inputs, smallholder organisation and access to markets. Some evidence was found of POs addressing the lack of credit. POs also often provide quality control and grading services, although the resources needed for this task often go beyond what the PO can bear. Extension services are inconclusive. One would expect technical assistance to become more important, particularly when quality requirements of final customers increase, however, POs often do not have the personnel and other resources to provide all members with good extension service.

Finally, we found strong evidence that partnerships address the lack of extension services and some evidence for the lack of organisation, quality control and inspection, and market linkages. The literature was inconclusive on the competences of partnerships in addressing the lack of financial services and market information. Moreover, we did not find empirical evidence that partnerships help smallholders overcome the lack of inputs.

When comparing the various institutional arrangements (by looking at the rows in the table), it appears that they often complement each other in terms of competences. Given these results, an investigation of how and under what conditions combinations of the three arrangements can be used effectively in quality improvement could bring interesting insights. The next section analyses more thoroughly that issue.

2.4 Interactions among and combinations of the different institutional arrangements

From Table 2.2 we can deduct, at least theoretically, that a combination of different institutional arrangements should be used to address all institutional constraints smallholder face when seeking to improve product quality. But what about empirical evidence? In the field, many institutional arrangements co-exist in coordinating activities along food chains. While reviewing the literature, we found evidence of interactions and combinations. Although not always directly linked to quality improvement, this section reports the main evidence of interactions found.

2.4.1 Contract farming combined with producer organisations

POs may engage in CF in order to support the efficiency and equity of the CF arrangement (Bijman and Wollni, 2009; Coulter *et al.*, 1999; Jia and Huang, 2011; Key and Runsten, 1999). At the efficiency level, Saenz-Segura *et al.* (2009) observe that POs can reduce the transaction costs in the contracting arrangement, as the contractor does not have to deal with numerous smallholder farmers but with only one organisation of smallholders. Okello and Swinton (2007) mention that exporters prefer to work with POs because they rationalise the costs of training and monitoring a multitude of small farmers. POs can also undertake the collection of products on behalf of the contractor. However, to justify group contracts, the costs of organising collection and transport and the costs of group membership need to be lower than the gains reached from lower rejection rates (Sáenz-Segura *et al.*, 2009). At the equity level, POs can improve the power balance between producers and contractors, thereby strengthening the incentives for both parties to continue bilateral contracting. POs can also serve as an enforcement mechanism between farmers and contracting firms by engaging their expertise in conflict resolution or by using common resources to hire such expertise. Finally, POs can participate in the elaboration of fair contract terms, for instance by designing standard contracts.

2.4.2 Contract farming combined with partnerships

Ensuring that production of smallholder farmers is aligned with market demand is easiest if leading exporters and purchasing firms are included in the institutional arrangement. In this manner, both partnerships and CF recognise the key role played by downstream market actors in integrating smallholders into high quality agrifood chains. Therefore, some partnerships have facilitated the establishment of CF relations or outgrower schemes between agribusiness and smallholder producers, which include the provision of extension services and credit to support the certification of smallholders (Bitzer and Glasbergen, 2010; Jaffee *et al.*, 2011). Boselie *et al.* (2003) note that there are effectively two roles of partnerships in CF. Firstly, partnerships may be essential to get a CF scheme off the ground, as donors can temporarily provide services, expertise and credit to farmers. In this way, partnerships can address the threshold transaction cost problem. Secondly, partnerships can improve the efficiency and/or counteract the potential negative effect of CF, for instance, by training farmers to bargain more effectively.

2.4.3 Partnerships combined with producer organisations

POs have a positive role to play in alleviating institutional constraints faced by smallholders farmers, but they seldom self-organise on a formal basis. Also, as mentioned earlier, they often lack financial as well as human resources. Therefore, external input is often needed to initiate POs and make them operational. The literature broadly distinguishes between three roles of partnerships in strengthening POs. Firstly, partnerships can provide capacity building by training group leaders and ensure that POs have sufficient marketing knowledge, management expertise and business skills (Narrod *et al.*, 2009; Perez-Aleman and Sandilands, 2008). Secondly, partnerships can induce changes in inter-farmer relationships and the forms of collective action. For instance, quality standards may require more strictly controlled membership and reduced group size to facilitate monitoring and training of farmers, which partnerships can support (Narrod *et al.*, 2009). Thirdly, partnerships can act as linking organisations, i.e. as (temporary) facilitators linking POs with other actors in the food chain (Bitzer *et al.*, 2011). For this role partnerships are better suited than government, given the wide evidence of rent seeking and elite capture in those farmer groups that have strong political affiliations (Narrod *et al.*, 2009).

An example of a successful collaboration of a PO with a partnership is given by Roy and Thorat (2008). The authors have reported on a public/private partnership marketing organisation that assists Indian grape farmer cooperatives. The marketing partner had two main roles: it acted as a facilitator to provide marketing expertise (negotiation of contracts, provision of information and supply certification) and it provided technical assistance and inputs to member farmers through the cooperatives.

2.5 Conclusions and policy implications

Recent trends in globalising markets, consumer demands, and retail strategies have made product quality one of the most important challenges for market access for smallholder farmers in developing countries. These farmers face a lack of appropriate resources and institutions both for helping them to improve quality and for connecting them to high-value markets. Farmers, therefore, still widely rely on cash-and-carry type of transactions or on informal institutions that only allow for inefficient trade practices. Institutional arrangements such as contracts, POs and public/private partnerships are increasingly being adopted in response to market and institutional failures. These institutional arrangements reduce the transaction costs that result from the need to strengthen coordination and to make specific investments in high-quality food chains.

The working and key success factors of these institutional arrangements have been reported in an increasing number of academic publications. Moreover, as these arrangements are gaining importance in practice, there are also high expectations among practitioners about their effectiveness. The objective of this paper was therefore to find evidence in the empirical literature that these arrangements indeed help smallholders overcome institutional constraints related to quality improvement, to compare the competences of each arrangement in addressing these constraints, and to discuss options for combining these arrangements.

After we identified the main institutional constraints faced by smallholders who seek to improve product quality, we compiled the empirical evidence on how the three institutional arrangements address those constraints. Our literature review demonstrates that the various institutional arrangements are often complementary. For instance, POs can be particularly useful to support the efficiency and equity of contracts. However, they often need an external promotor to become operational and effective. Partnerships can be complementary to POs in providing for human, technical and financial resources. More generally, partnerships seem to be especially well suited in the initial phase of setting up collective action or CF schemes, especially for financial support, technical assistance and training.

Although our knowledge of the effect of innovative institutional arrangements on quality is increasing with the publication of enlightening studies, this chapter shows that there are still many knowledge gaps. We first need to deepen our understanding of the potential interaction between each institutional arrangement. For instance, Arinloye *et al.* (2012) showed that farmers involved in outgrowing (CF) schemes are less likely to be involved in other institutional arrangements, showing the specificity and exclusivity of this type of institutional arrangement. Another gap concerns the role of POs and partnerships in mitigating the drawbacks of CF. The functioning and impact of this 'double hybrid' institutional arrangement on quality outcomes, although increasingly used in developed countries, has hardly been addressed. Finally, from a more macro-level perspective, we need to assess the effect of the formal institutional environment on quality upgrading and on the development and efficiency of institutional arrangements. Studies comparing the same crop among different countries having different institutions could shed lights on that issue.

References

Altenburg, T., 2007. Donor approaches to supporting pro-poor value chains. Report prepared for the Donor Committee for Enterprise Development Working Group on Linkages and Value Chains. German Development Institute (DIE), Bonn, Germany.

Arinloye, D.D.A.A., Hagelaar, G., Linnemann, A.R., Pascucci, S., Coulibaly, O., Omta, S.W.F. and Van Boekel, M.A.J.S. (2012). Multi-governance choices by smallholder farmers in the pineapple supply chain in Benin: an application of transaction cost theory. African Journal of Business Management 6: 10320-10331.

Asfaw, S., Mithöfer, D. and Waibel, H., 2010. What impact are EU supermarket standards having on developing countries' export of high-value horticultural products? Evidence From Kenya. Journal of International Food and Agribusiness Marketing 22: 252-276.

Barrett, C.B., 2008. Smallholder market participation: concepts and evidence from eastern and southern Africa. Food Policy 33: 299-317.

Berdegué, J.A., Balsevich, F., Flores, L. and Reardon, T., 2005. Central American supermarkets' private standards of quality and safety in procurement of fresh fruits and vegetables. Food Policy 30: 254-269.

Biénabe, E. and Sautier, D., 2005. The role of small scale producers' organizations to address market access. Available at: http://tinyurl.com/hb3xxmw.

Bijman, J. and Wollni, M., 2009. Producer organisations and vertical coordination: an economic organization theory perspective. In: Rösner, H.J. and Schulz-Nieswandt, F. (eds.) Beiträge der genossenschaftlichen Selbsthilfe zur wirtschaftlichen und sozialen Entwicklung. LIT Verlag, Berlin, Germany, pp. 231-252.

Bijman, J., Muradian, R. and Schuurman, J. (eds.), 2016. Cooperatives, economic democratization and rural development. Edward Elgar, Cheltenham, UK.

Bitzer, V. and Glasbergen, P., 2010. Partnerships for sustainable change in cotton: an institutional analysis of African cases. Journal of Business Ethics 93: 223-240.

Bitzer, V., Francken, M. and Glasbergen, P., 2008. Intersectoral partnerships for a sustainable coffee chain: really addressing sustainability or just picking (coffee) cherries? Global Environmental Change 18: 271-284.

Bitzer, V., Van Wijk, J., Helmsing, A.H.J. and Van der Linden, V., 2011. Partnering to facilitate smallholder inclusion in value chains. In: Helmsing, A.H.J. and Vellema, S. (eds.) Value chains, social inclusion and economic development: contrasting theories and realities. Routledge, London, UK, pp. 221-246.

Boselie, D., Henson, S. and Weatherspoon, D., 2003. Supermarket procurement practices in developing countries: redefining the roles of the public and private sectors. American Journal of Agricultural Economics 85: 1155-1161.

Brester, G.W. and Penn, J.B., 1999. Strategic business management principles for the agricultural production sector in a changing global food system. Trade Research Center, Montana State University, Bozeman, MT, USA.

Cechin, A., Bijman, J., Pascucci, S., Zylbersztajn, D. and Omta, O., 2013. Quality in cooperatives versus investor-owned firms: evidence from broiler production in Paraná, Brazil. Managerial and Decision Economics 34: 230-243.

Chitundu, M., Droppelmann, K. and Haggblade, S., 2009. Intervening in value chains: lessons from Zambia's task force on acceleration of cassava utilisation. Journal of Development Studies 45: 593-620.

Cook, M.L., 1995. The future of U.S. agricultural cooperatives: a neo-institutional approach. American Journal of Agricultural Economics 77: 1153-1159.

Coulter, J., Goodland, A., Tallontire, A. and Stringfellow, R., 1999. Marrying farmer cooperation and contract farming for service provision in a liberalizing Sub-Saharan Africa. Natural Resources Perspectives # 48. Overseas Development Institute (ODI), London, UK.

Davis, L. and North, D.C., 1971. Institutional change and american economic growth. Cambridge University Press, Cambridge, UK.

Donovan, J., Franzel, S., Cunha, M., Gyau, A. and Mithöfer, D., 2015. Guides for value chain development: a comparative review. Journal of Agribusiness in Developing and Emerging Economies 5: 2-23.

Dorward, A., Fan, S., Kydd, J., Lofgren, H., Morrison, J., Poulton, C., Rao, N., Smith, L., Tchale, H., Thorat, S., Urey, I. and Wobst, P., 2004. Institutions and policies for pro-poor agricultural growth. Development Policy Review 22: 611-622.

Dorward, A., Kydd, J., Morrison, J. and Poulton, C., 2005. Institutions, markets and economic coordination: linking development policy to theory and praxis. Development and Change 36: 1-25.

Dorward, A., Kydd, J., Poulton, C., 1998. Smallholder cash crop production under market liberalisation. a new institutional economics perspective. CAB International, Wallingford, UK.

Dorward, A., Poole, N., Morrison, J., Kydd, J. and Urey, I., 2003. Markets, institutions and technology: missing links in livelihoods analysis. Development Policy Review 21: 319-332.

Eaton, C. and Shepherd, A.W., 2001. Contract farming. Partnerships for growth. FAO Agricultural Services Bulletin.

Fafchamps, M. and Hill, R., 2005. Selling at the farm-gate or travelling to market. American Journal of Agricultural Economics 87: 717-734.

Ferroni, M. and Castle, P., 2011. Public-private partnerships and sustainable agricultural development. Sustainability 3: 1064-1073.

Francesconi, G.N., Heerink, N. and D'haese, M., 2010. Evolution and challenges of dairy supply chains: evidence from supermarkets, industries and consumers in Ethiopia. Food Policy 35: 60-68.

Fuchs, D., Kalfagianni, A. and Havinga, T., 2011. Actors in private food governance: the legitimacy of retail standards and multistakeholder initiatives with civil society participation. Agriculture and Human Values 28: 353-367.

Gabre-Madhin, E., 2009. A market for all farmers: market institutions and smallholder participation. University of California, Berkeley, CA, USA.

Gereffi, G., 1999. International trade and industrial upgrading in the apparel commodity chain. Journal of International Economics 48: 37-70.

Gereffi, G., Humphrey, J. and Sturgeon, T., 2005. The governance of global value chains. Review of International Political Economy 12: 78-104.

Glasbergen, P., 2007. Setting the scene: the partnership paradigm in the making. In: Glasbergen, P., Biermann, F. and Mol, A.P.J. (eds.) Partnerships, governance and sustainable development. Reflections on theory and practice. Edward Elgar, Cheltenham, UK, pp. 1-28.

Glover, D.J., 1984. Contract farming and smallholder outgrower schemes in less-developed countries. World Development 12: 1143-1157.

Glover, D.J. and Kusterer, K., 1990. Small farmers, big business: contract farming and rural development. Macmillan, London, UK.

Goldsmith, A., 1985. The private sector and rural development: can agribusiness help the small farmer? World Development 13: 1125-1138.

Haantuba, H. and De Graaf, J., 2008. Linkages between smallholder farmers and supermarkets: lessons from Zambia. In: McCullough, E.B., Pingali, P.L. and Stamoulis, K.G. (eds.) The transformation of agri-food systems: globalization, supply chains and smallholder farmers. Earthscan, London, UK, pp. 207-226.

Hatanaka, M., Bain, C. and Busch, L., 2005. Third-party certification in the global agrifood system. Food Policy 30: 354-369.

Hellin, J., Lundy, M. and Meijer, M., 2009. Farmer organization, collective action and market access in Meso-America. Food Policy 34: 16-22.

Henson, S. and Reardon, T., 2005. Private food safety and quality standards for fresh produce exporters: the case of Hortico Agrisystems, Zimbabwe. Food Policy 30:371-384.

Henson, S., Masakure, O. and Cranfield, J., 2011. Do fresh produce exporters in Sub-Saharan Africa benefit from GlobalGAP certification? World Development 39: 375-386.

Holloway, G., Nicholson, C., Delgado, C., Staal, S. and Ehui, S., 2000. Agroindustrialization through institutional innovation. Transaction costs, cooperatives and milk-market development in the east-African highlands. Agricultural Economics 23: 279-288.

Hueth, B., Ligon, E., Wolf, S. and Wu, S., 1999. Incentive instruments in fruit and vegetable contracts: input control, monitoring, measuring, and price risk. Review of Agricultural Economics: 374-389.

Humphrey, J. and Navas-Alemán, L., 2010. Value chains, donor interventions and poverty reduction: a review of donor practice. Research report 63. IDS, Brighton, UK. Available at: http://tinyurl.com/jofjofy.

Jaffee, S. and Henson, S., 2005. Agro-food exports from developing countries: the challenges posed by standards. In: Aksoy, A. and Beghin, J.C. (eds.) Global agricultural trade and developing countries. World Bank, Washington, DC, USA, pp. 91-114.

Jaffee, S. and Masakure, O., 2005. Strategic use of private standards to enhance international competitiveness: vegetable exports from Kenya and elsewhere. Food Policy 30: 316-333.

Jaffee, S., Henson, S. and Díaz Rios, L., 2011. Making the grade: smallholder farmers, emerging standards, and development assistance programs in Africa (a research program synthesis), World Bank, Washington, DC, USA.

Jayne, T.S., Mather, D. and Mghenyi, E., 2010. Principal challenges confronting smallholder agriculture in Sub-Saharan Africa. World Development 38: 1384-1398.

Jia, X. and Bijman, J., 2014. Contract farming: synthetic themes for linking farmers to demanding markets. In: Da Silva, C.A. and Rankin, M. (eds.) Contract farming for inclusive market access. FAO, Rome, Italy, pp. 21-38.

Jia, X. and Huang, J., 2011. Contractual arrangements between farmer cooperatives and buyers in China. Food Policy 36: 656-666.

Kaganzi, E., Ferris, S., Barham, J., Abenakyo, A., Sanginga, P.C. and Njuki, J., 2009. Sustaining linkages to high value markets through collective action in Uganda. Food Policy 34: 23-30.

Kaplinsky, R., 2000. Spreading the gains from globalisation: what can be learned from value chain analysis? Journal of Development Studies 37: 117-146.

Kersting, S. and Wollni, M., 2011. Public-private partnerships and GLOBALGAP standard adoption: evidence from small-scale fruit and vegetable farmers in Thailand. In: EAAE 2011 Congress Change and Uncertainty, Zurich.

Key, N. and Runsten, D., 1999. Contract farming, smallholders, and rural development in Latin America: the organization of agroprocessing firms and the scale of outgrower production. World Development 27: 381-401.

Kindness, H. and Gordon, A., 2001. Agricultural marketing in developing countries: the role of NGOs and CBOs. University of Greenwich, Greenwich, UK.

Kirsten, J.F., Dorward, A., Poulton, C. and Vink, N., 2009. Institutional economic perspectives on african agricultural development. International Food Policy Research Institute (IFPRI), Washington, DC, USA.

Kolk, A., Van Tulder, R. and Kostwinder, E., 2008. Business and partnerships for development. European Management Journal 26: 262-273.

Kydd, J. and Dorward, A., 2001. The Washington consensus on poor country agriculture: analysis, prescription and institutional gaps. Development Policy Review 19: 467-478.

Kydd, J. and Dorward, A., 2004. Implications of market and coordination failures for rural development in least developed countries. Journal of International Development 16: 951-970.

Little, P.D. and Watts, M.J., 1994. Living under contract: contract farming and agrarian transformation in Sub-Saharan Africa. University of Wisconsin Press, Madison, WI, USA.

Maertens, M. and Swinnen, J.F.M., 2009. Trade, standards, and poverty: evidence from Senegal. World Development 37: 161-178.

Markelova, H., Meinzen-Dick, R.S., Hellin, J. and Dohrn, S., 2009. Collective action for smallholder market access. Food Policy 34: 1-7.

Minot, N.W., 1986. Contract farming and its effects on small farmers in less developed countries. MSU International Development Papers. Michigan State University, Department of Agricultural Economics, East Lansing, MI, USA.

Minten, B., Randrianarison, L. and Swinnen, J.F.M., 2009. Global retail chains and poor farmers: evidence from Madagascar. World Development 37: 1728-1741.

Moustier, P., Tam, P.T.G., Anh, D.T., Binh, V.T. and Loc, N.T.T. (2010). The role of farmer organizations in supplying supermarkets with quality food in Vietnam. Food Policy 35: 69-78.

Muriithi, B.W., Mburu, J. and Ngigi, M., 2011. Constraints and determinants of compliance with EurepGap standards: a case of smallholder french bean exporters in Kirinyaga district, Kenya. Agribusiness 27: 193-204.

Narrod, C., Devesh, R., Okello, J., Avendaño, B., Rich, K. and Thorat, A., 2009. Public-private partnerships and collective action in high value fruit and vegetable supply chains. Food Policy 34: 8-15.

North, D.C., 1990. Institutions, institutional change and economic performance. Cambridge University Press, Cambridge, UK.

Okello, J. and Swinton, S.M., 2007. Compliance with international food safety standards in Kenya's green bean industry: comparison of a Smalland a large-scale farm producing for export. Review of Agricultural Economics 29: 269-285.

Penrose-Buckley, C., 2007. Producer organisations. a guide to developing collective rural enterprises. Oxfam GB, Oxford, UK.

Perez-Aleman, P. and Sandilands, M., 2008. Building value at the top and the bottom of the global supply chain: MNC-NGO partnerships. California Management Review 51: 24-49.

Poulton, C., Dorward, A. and Kydd, J., 2010. The future of small farms: new directions for services, institutions, and intermediation. World Development 38: 1413-1428.

Poulton, C., Kydd, J. and Dorward, A., 2006. Overcoming market constraints on pro-poor agricultural growth in Sub-Saharan Africa. Development Policy Review 24: 243-277.

Raynaud, E. and L. Sauvée (2000). Common labelling and producer organisations: a transaction cost economics approach. In: Sylvander, B., Barjolle, D. and Arfini, F. (eds.) The socio economics of origin labelled products in agri-food supply chains: spatial, institutional and co-ordination aspects. INRA Actes et Communications Vol. 17, pp. 133-142.

Raynolds, L.T., 2004. The globalization of organic agro-food networks. World Development 32: 725-743.

Reardon, T. and Barrett, C.B., 2000. Agroindustrialization, globalization, and international development: an overview of issues, patterns, and determinants. Agricultural Economics 23: 195-205.

Reardon, T., Timmer, C.P., Barrett, C.B. and Berdegué, J., 2003. The rise of supermarkets in Africa, Asia, and Latin America. American Journal of Agricultural Economics 85: 1140-1146.

Rivera-Santos, M., Rufín, C. and Kolk, A., 2002. Bridging the institutional divide: partnerships in subsistence markets. Journal of Business Research 65: 1721-1727.

Rondot, P. and Collion, M.-H., 2001. Agricultural producers organizations: their contribution to rural capacity building and pverty reduction. World Bank, Washington, DC, USA.

Roy, D. and Thorat, A., 2008. Success in high value horticultural export markets for the small farmers: the case of Mahagrapes in India. World Development 36: 1874-1890.

Sáenz-Segura, F., D'Haese, M. and Schipper, R.A., 2009. A seasonal model of contracts between a monopsonistic processor and smallholder pepper producers in Costa Rica. Agricultural Systems 103: 10-20.

Schäferhoff, M., Campe, S. and Kaan, C., 2009. Transnational public-private partnerships in international relations: making sense of concepts, research frameworks, and results. International Studies Review 11: 451-474.

Schouten, G. and Glasbergen, P., 2011. Creating legitimacy in global private governance: the case of the roundtable on sustainable palm oil. Ecological Economics 70: 1891-1899.

Shepherd, A.W., 2007. Approaches to linking producers to markets. A review of experiences to date. FAO, Rome, Italy.

Shiferaw, B., Obare, G. and Muricho, G., 2006. rural institutions and producer organizations in imperfect markets: experiences from producer marketing groups in semi-arid Eastern Kenya. International Crops Research Institute for the Semi-Arid Tropics, Patancheru, Andhra Pradesh, India.

Singh, S., 2002. Contracting out solutions: political economy of contract farming in the Indian Punjab. World Development 30: 1621-1638.

Spielman, D.J., Byerlee, D., Alemu, D. and Kelemework, D., 2010. Policies to promote cereal intensification in Ethiopia: the search for appropriate public and private roles. Food Policy 35: 185-194.

Springer-Heinze, A., 2007. ValueLinks manual – the methodology of value chain promotion. German Technical Cooperation (GTZ), Eschborn, Germany.

Swinnen, J.F.M. and Maertens, M., 2007. Globalization, privatization, and vertical coordination in food value chains in developing and transition countries. Agricultural Economics 37 Suppl. 1: 89-102.

Sykuta, M.E. and Cook, M.L., 2001. A new institutional economics approach to contracts and cooperatives. American Journal of Agricultural Economics 83: 1273-1279.

Thiele, G., Devaux, A., Reinoso, I., Pico, H., Montesdeoca, F., Pumisacho, M., Andrade-Piedra, J., Velasco, C., Flores, P., Esprella, R., Thomann, A., Manrique, K. and Horton, D., 2011. Multi-stakeholder platforms for linking small farmers to value chains: evidence from the Andes. International Journal of Agricultural Sustainability 9: 423-433.

Thorp, R. Stewart, F. and Heyer, A., 2005. When and how far is group formation a route out of chronic poverty? World Development 33: 907-920.

Trienekens, J.H. 2011. Agricultural value chains in developing countries; a framework for analysis. International Food and Agribusiness Management Review 14, 51-83.

Van Tulder, R. and Fortanier, F., 2009. Business and sustainable development: from passive involvement to active partnerships. In: Kremer, M., Van Lieshout, P. and Went, R. (eds.) Doing good or doing better: development policies in a globalizing world. Amsterdam University Press, Amsterdam, the Netherlands, pp. 211-235.

Van Wijk, J. and Kwakkenbos, H., 2012. Beer multinationals Supporting Africa's Development? How partnerships include smallholders into sorghum-beer supply chains. In: Van Dijk, M.P. and Trienekens, J.H. (eds.) Global value chains: linking local producers from developing countries to international markets. Amsterdam University Press, Amsterdam, the Netherlands, pp. 71-88.

Verhaegen, I. and Van Huylenbroeck, G., 2002. Hybrid governance structures for quality farm products: a transaction cost perspective. Shaker Verlag, Aachen, Germany.

Vorley, B., Fearne, A. and Ray, D., 2007. Regoverning markets: a place for small-scale producers in modern agrifood chains? Aldershot, Burlington, UK.

Weatherspoon, D. and Reardon, T., 2003. The rise of supermarkets in Africa: implications for agrifood systems and the rural poor. Development Policy Review 21: 333-355.

Winfree, J.A. and McCluskey, J.J., 2005. Collective reputation and quality. American Journal of Agricultural Economics 87: 206-213.

Wolf, S., Hueth, B. and Ligon, E., 2001. Policing mechanisms in agricultural contracts. Rural Sociology 66: 359-381.

World Bank, 2007. World Development Report 2008. Agriculture for development. World Bank, Washington, DC, USA.

3. Quality challenges and opportunities in the pineapple supply chain in Benin

D.D.A.A. Arinloye[1]* and M.A.J.S. van Boekel[2]

[1]*World Agroforestry Centre (ICRAF), Bamako Mali, BPE5118 Bamako, Mali;* [2]*Wageningen University, Food Quality and Design Group, Agrotechnology and Food Sciences, Bornse Weilanden 9, 6708 WG Wageningen, the Netherlands; a.arinloye@cgiar.org*

Abstract

Quality has become a key aspect for establishing international market access and improving competitiveness of (smallholder) producers in developing countries. This is especially the case for perishable tropical food products. This chapter explores the key quality issues of the Beninese pineapple sector. An analysis is made of the constraints and opportunities for improving pineapple quality at production, processing, trading and post-harvest levels. The chapter concludes with recommendations on how various actors in the institutional environment of Benin, such as governmental agencies and NGOs, can contribute to quality improvements of the pineapple sector.

Keywords: quality, value chain, pineapple, Benin

3.1 Introduction

Quality has become an important parameter in the production and marketing of food products. Food quality is a difficult concept, it is actually a property assigned by consumers to a product and there is, therefore, an interplay between product properties and consumer (Van Boekel, 2009). There are subjective and objective aspects to quality. For instance, safety and nutritional value can be considered as objective quality attributes (but can only be measured in a laboratory and cannot be judged by consumers) while taste and flavour are subjective (but are difficult to measure analytically). Yet, there are rules and regulations concerning food quality (mostly related to food safety) that need to be obeyed, sometimes nationally and increasingly internationally. Small-scale farmers, particularly those in sub-Saharan Africa, often face difficulties in meeting these quality standards. This can be due to a variety of reasons including lack of knowledge, lack of resources (capital investments, fertiliser, herbicides, pesticides, etc.), lack of infrastructure and suitable transport, and lack of institutional support (Ruben *et al.*, 2007; Trienekens and Zuurbier, 2008; Ziggers and Trienekens, 1999).

Safety, quality, health, sustainability and integral quality care and supply chain management have become key features for market access and competitiveness. Producers and processors of food products need to respond to domestic and international market challenges related to rapid urbanisation and demographic change in the developing world, increasing participation of supermarkets in food provision, and globalisation of food supply chains and networks. The expanding international markets offer opportunities but also many challenges for small-scale producers and value chain actors in less developed and developing countries. Quality challenges are among major barriers to supply chain development. These quality challenges are even pronounced for supply chains of vulnerable and perishable tropical products that involve small-scale producers from developing countries.

Pineapple (*Ananas comosus*) is one of the tropical perishable products gaining attention and interest of policies and chain actors in many sub-Saharan African countries. Pineapple is the third most important tropical fruit in West Africa after plantains and citrus fruits (Faostat, 2012). In Benin, pineapple is promoted by the Beninese government as one of the key products for exports to become less dependent on a single crop, cotton. Pineapple production areas are mainly based in Southern Benin. Although cultivated between Abomey and Savè, pineapple is most successfully produced on tray bar soils. These are the plates of Allada, Abomey, Pobè, Kétou Sakété, and Aplahoué. The largest area of pineapple production is located in the department of Atlantic. Two varieties of pineapple are produced: the 'Smooth cayenne', which is destined for export markets, and 'Sugarloaf' which mainly sold in local and regional markets (Fassinou *et al.*, 2012). In 2009, there were approximately 10,000 pineapple growers in Benin (Dohou, 2008). Pineapple production has grown rapidly over the past decade, and it now represents the third largest export product of the country, after cotton and cashew (ITC, 2013). This rapid growth is due to both a strong increase of local and regional demand and a relative success of a few agricultural and trade entrepreneurs in serving European markets. However, getting high pineapple quality remains problematic for most small-scale producers that wish to make benefits of lucrative export prices (Mongbo and Floquet, 2006).

The main problem of pineapple production in Benin is the poor quality and this applies to all the major markets, i.e. local, regional and European markets (Fassinou *et al.*, 2012; Gbenou *et al.*, 2006). Small-scale producers face important barriers to improving quality. This leads to heterogeneous quality of their products that is often below domestic and international market requirements (Dolan and Humphrey, 2000; Trienekens and Zuurbier, 2008). To access export markets, small-scale producers have to meet high product quality demands by complying with market requirements based on standards and regulations. There, producers also have to cope with a lack of institutional and infrastructural support, availability of resources and efficient and effective coordination along the value chains. It is then important to look for ways to improve quality in production and processing. It is proposed that this can only be obtained in the organisation of the supply chain, when technical innovations in product and process are combined with support from adequate institutions that will make technical improvements efficient (Dorward *et al.,* 2004).

The present chapter aims to shed light on the key quality issues of a perishable product like pineapple in Benin and to identify opportunities to improve quality. The next section describes the quality issues at producer, processor, trader, and consumers' levels. Section 3.3 shows opportunities for quality improvement and in Section 3.4 we draw some conclusions.

3.2 Quality issues along the pineapple value chain

Nowadays, quality is a very important issue in the marketing of food products (Henson and Reardon, 2005; Reardon and Farina, 2002; Royer and Bijman, 2012). Quality problems are due to a variety of reasons, as mentioned in the introduction. Small-scale producers usually have limited access to good quality planting materials; have poor storage and processing facilities, have little knowledge about how to influence quality. This section presents common issues along the Beninese pineapple supply chain at fruit production, processing, trading, post-harvesting, and consumption levels (Appendix 3.1 and 3.2).

3.2.1 Fruit production

The pineapple production cycle from planting to harvest lasts 15 to 18 months on average. The mean annual rainfall favourable for pineapple growth and development is 1,200 mm, optimum rainfall for good commercial pineapple cultivation ranges from 1000 mm to 1,500 mm (Bartholomew *et al.,* 2003). As for soil characteristics, good drainage and pH ranging from 5.5-6.0 are favourable as well; the best soils for pineapple culture have a neutral to acid pH (Hepton, 2003; Morton, 1987) with good drainage (Collins, 1968; Hepton, 2003) in order to prevent water logging and root diseases. The main constraints noticed with pineapple producers are: (1) non-availability of planting material (mainly for the 'Smooth cayenne' cultivar); (2) heterogeneity of planting material; (3) variation in planting material age; (4) non-availability of fertilisers (mainly the sulphate of potash K_2SO_4); (5) high susceptibility of Smooth cayenne to the wilt disease; and (6) neglecting the Ethephon residue in the fruit at harvest time (Fassinou *et al.,* 2012). These constraints may be related to cultural practices (Brown, 1986).

The cultural practices of 'Smooth cayenne' and 'Sugarloaf' are different. The first step of planting is land preparation. Fassinou *et al.* (2012) have noticed that producers preferred the beginning of the first rainy season as planting time and planting materials used included all traditional types of planting materials: slips (produced on the peduncle at the base of the fruit), *hapas* or side shoots, and suckers (preferentially for 'Sugarloaf'), while only *hapas* and suckers were used by 'Smooth cayenne' producers. It has been observed that crowns were not used. Planting materials were either from plants kept in the field after the previous harvest, or from other producers, and for some cases, they can be from both (Fassinou *et al.*, 2012). Heterogeneity in planting material is in weight, age and in number of leaves and this could contribute to the heterogeneity in pineapple quality since there is a relation between the size and type of planting material and fruit size (Linford *et al.*, 1934; Malézieux *et al.*, 1994). Furthermore, it has been shown that the quality of the planting material is a determinant of the quality of the final product. The more viable and healthy the planting material, the better the quality of the fruits (Arinloye, 2013; Garnier, 1997).

3.2.2 Processing

Pineapple can be processed in several ways. It is one of the most popular fruits used to make wine in many developing countries (FAO, 1997), but this is not yet well developed in West African countries. In Benin, pineapple is transformed into juice or concentrated juice, or dehydrated pineapple. The juice or concentrated pineapple juice is one of the most consumed tropical fruits juices. It represents about 5% of the Beninese fruit juice market which is dominated by orange juice, apple and grapes (E.G.S Adjovi, personal communication). About 15% of produced fresh pineapple is processed into juice exclusively sold in local and regional markets (Arinloye, 2013). Processors obtain their fruits from wholesalers or directly from producers because of the quality requirements in processing pineapples. These are: the big size of the fruit, high sugar content and no skin damages (Royer and Bijman, 2012). Dehydrated pineapple comprises 5,000 tons and the main outlets are Germany, United Kingdom, the Netherlands, Japan and the United States. Also, there are frozen pineapple slices and pieces, but this value chain is still limited to the few supermarkets and restaurants. Furthermore, the use of pineapple seems to be growing in other products; it is used in pastry, dairy products, baby' food and fruit salads.

In general, export of processed pineapple from Benin is still marginal, mainly because of difficulties to obtain sufficient and homogeneous quality (Royer and Bijman, 2012). Finding suitable pineapples on the market to process pineapple at a consistent high-quality level remains a major challenge for processors (Hounhouigan, 2014). Quality of processed pineapple can also be affected by processing techniques. Pasteurisation, which is a process of heating a food to a specific temperature for a predefined length of time and then immediately cooling it after it is removed from the heat (FAO, 1997), is one these processing techniques. Although pasteurisation is needed for preventing spoilage, it also induces the so-called Maillard reaction, leading to brown discolouration of the juice, especially when stored at relatively high temperature. This browning reaction limits shelf life considerably and is an obstacle for export. Furthermore, a small loss of vitamin C may occur due to pasteurisation. To the best of our knowledge, no study is done so far on the quality of the pasteurised pineapple juice in Benin, however research is underway (Hounhouigan, 2014). It is therefore important to improve processing practices by using modern and hygienic equipment (Adossou, 2012).

3.2.3 Trading

Fresh pineapples from Benin are sold on domestic, regional and European markets (Fassinou *et al.*, 2012; Gbenou *et al.*, 2006). The domestic and regional markets are generally characterised by lower (or non-existent) quality and standards requirements while the European markets are characterised by higher quality and standards requirements (Arinloye, 2013). In Europe, the market for fresh pineapples is one of the fastest growing fruit and vegetable markets in Europe (Pay, 2009). Europe imports annually approximately 500,000 tons of fresh pineapple (INSAE, 2009). Costa Rica is by far the largest exporter of pineapples to the European market, supplying 73% of all imports in 2008. As for Benin, less than 1% of its production is exported to Europe annually. Although the taste of the pineapple produced in Benin is nicely appreciated on the global market (E.G.S Adjovi, personal communication), complying with international sanitary and phytosanitary standards remains a big challenge for both producers and marketers. Beninese fresh pineapple export must compete with other regional countries such as Côte d'Ivoire, Cameroun and Ghana on the European market (Royer and Bijman, 2012). The fact that Beninese pineapples do not meet European quality norms might thus play an important role in the slow progress in exports. These quality norms include the *Codex Alimentarius*, safety norms on maximum residue levels, heavy metals maximum levels, GlobalGAP regulations and traceability norms (Daniel and Martin, 2008; Royer and Bijman, 2012).

The local market is, therefore, the most important outlet for Benin pineapple although the proportion of pineapple sold on that market is not well-known; it varies from 26 to 60% depending on the sources consulted (Dohou, 2008; Helvetas-Benin, 2007).

3.2.4 Post-harvest practices

Post-harvest handling is very important for pineapple quality. Standards of sanitation are needed to comply with most export food safety standards (Jaffee and Henson, 2004). Exporters mentioned that the absence of refrigerated storage facilities at the airport affected pineapple quality (Royer and Bijman, 2012). This increases fresh pineapple losses up to about 25% during transport (Adossou, 2012). Other factors have been argued to have negative effects on pineapple quality, such as pineapple overloading in cars or trucks during transportation from farms to markets that causes pineapples at the bottom to be damaged and/or spoiled, and the use of cars or trucks previously used to transport chemicals or fertilisers. Also, pineapple, after harvest, should not be left exposed to direct sunlight; this affects its quality considerably (Batholomew *et al.*, 2003).

3.2.5 Consumption

The role of fruits and fruit juices in nutrition and health is very important. More and more, consumers look for better quality of fruits with regards to certain characteristics related to the quality (Wardy *et al.*, 2009). Pineapple is one of the most popular of the non-citrus tropical and subtropical fruits, because of its attractive flavour and refreshing sugar-acid balance (Bartolomé et al., 1995). Pineapple juice is consumed worldwide, mainly in reconstituted or concentrated form and in the mixture composition to obtain new flavours in cocktails and other products (Arthey, 1995; De Carvalho et al., 2008; Jan and Masih, 2012). Pineapple mixture composition has been investigated in Ghana by Wardy et al. (2009) and comparative studies have been made on the three main pineapple varieties ('Smooth cayenne', 'Sugarloaf' and 'MD2') cultivated, based on physical, chemical and sensory aspects. They found that 'Sugarloaf' had the highest juice volume of 206 ml/kg fruit, followed by 'MD2' (134-191 ml/kg) with 'Smooth cayenne' having the smallest volume. They also found that there were no significant flavour differences; although there were significant differences in the overall preference for fruit juices indicating that 'MD2' was the most preferred pineapple fruit. The ascorbic acid content of pineapple is variable depending on factors such as the cultivar, stage of maturity, conditions of storage and the part of fruit (Collins, 1968; Ngoddy and Ihekoronye, 1985). The vitamin C content of the 'MD2', 'Sugarloaf' and 'Smooth cayenne' meets the dietary requirement for vitamin C, but 'MD2' has the best potential due to its very high levels on weight (Wardy et al., 2009). As a negative quality aspect, the vitamin C amount in a fruit has been linked with internal browning associated with post-harvest chilling injury (Miller, 1951), though the main reason for browning in juice is the already mentioned Maillard reaction.

3.3. Opportunities for quality improvement

Quality improvement along the pineapple chain involves improved agronomic, and technological management practices, as well as institutional considerations. This review focuses on the latter aspect. Pineapple chain actors generally lack logistic and infrastructure supports (Arinloye, 2013), the legal framework is missing, and one is faced with significant contract enforcement issues. Logistic and infrastructure supports include market information, basic information and transportation infrastructures, and storage facilities. Lack of market information and basic information infrastructure may lead to information asymmetry between pineapple producers and others actors. Lack of proper transportation infrastructure may increase transportation costs, and thereby transaction costs. Lack of storage facilities results in spoiling of fresh pineapples. Providing logistic and infrastructure supports may therefore facilitate quality improvement along the chain. For example in Africa, the use of mobile phone is widespread (e.g. Gray, 2006), with more than 87% of small-scale farmers in rural areas in Benin using the technology (Arinloye, 2013). Mobile technology has facilitated farmers and many entrepreneurs' linkage and easier access to markets (Bertolini, 2004), it constitutes an important opportunity to build on quality improvement strategies.

The legal framework includes (international) laws that regulate product quality in the domain. It prescribes requirements for high quality products, including quality norms and standards, measures based on the principles of risk analysis. The adoption and implementation of legislative measures that comply with the importing countries' quality regulations may be the first step to encourage quality improvement along the pineapple chain. Sensitising and training producers and others actors in quality norms and standards, and providing laboratories with adequate equipment to perform quality testing (Adossou, 2012) may constitute the second step towards quality improvement.

Contract represents forward agreements, such as the volumes and quality of products, prices and payment systems, defined by trading parties to control their transactions (e.g. Dorward *et al.*, 2004). It serves not only as a means of keeping transaction costs low (Williamson, 1979), but also as an incentive for trading parties to, for example, invest in quality, as they constitute a safeguard against opportunism (e.g. Gibbons, 2005; Williamson, 1991). The enforcement of contracts between small-scale producers and other chain actors in developing countries is often subject to conflicts. This increases the perceived risks of transacting and thereby transaction costs in that part of the world, as low-cost and reliable formal contract enforcement institutions often have weaker influence (e.g. Ingenbleek *et al.*, 2013). Instead, the organisation of farmers into group actions such as cooperatives, is more generalised. Many studies point out the roles of farmer-based organisations in guaranteeing contract enforcement in developing countries. For example, McMillan and Woodruff (2000) found in Vietnam and Eastern Europe that trade associations, besides their primary role as information providers, also supplement courts in facilitating dispute resolution and arbitration. Promoting and strengthening small-scale producers' cooperatives/associations and chain member cooperatives may therefore be an effective way to encourage quality improvement along the pineapple chain.

3.4 Conclusions

The aim of the present chapter was to discuss quality issues and quality improvement opportunities along the pineapple chain in Benin. Although pineapple production has become increasingly important over the past years, quality issues remain problematic along the value chains, with producers facing significant challenges in meeting export market quality requirements. The present book chapter highlights opportunities through which product quality can be improved. Appendix 3.1 gives a SWOT analysis for the pineapple chain while Appendix 3.2 lists the constraints and possible solutions. Market information in particular about quality requirements should be provided to producers. The collection and dissemination of such information has to be ensured by relevant chain institutions (e.g. extension services, non-governmental organisations and/or development projects), as well as by exporters themselves. The collection and dissemination of market information to producers by exporters can be encouraged by the promotion of contract farming between actors, in that it might increase the level of trust between trading partners and by that facilitating mutual assistance such as pre-financing credit to producers. An improved use of mobile phones in agriculture development, including the Beninese pineapple supply chains, can also be used to strengthen market information sharing among producers. Policy makers, non-governmental organisations and development projects that aim at strengthening the capacity of producers to improve the quality level of their produce have to upgrade their legislative measures to comply with the importing countries' laws and regulations, and provide actors in the chain with infrastructure supports and training on quality norms and standards.

References

Adossou, F.T., 2012. Introduction to pineapple industry in Benin. Pineapple News 19: 6-14.

Arinloye, D.D.A.A., 2013. Governance, marketing and innovations in Beninese pineapple supply chains: a survey of smallholder farmers in South Benin. PhD thesis, Wageningen University, Wageningen, the Netherlands.

Arthey, D., 1995. Food industries manual. In: Ranken, M.D., Kill, R.C. and Baker, C. (eds.) Fruit and vegetable product. Blackie Academic and Professional, London, UK, pp. 151-175.

Bartolomé, A.P., Rupérez, P. and Fúster, C., 1995. Pineapple fruit: morphological characteristics, chemical composition and sensory analysis of red Spanish and smooth cayenne cultivars. Food Chemistry 53: 75-79.

Bartholomew, D.P., Paull, R.E. and Rohrbach, K.G., 2003. The pineapple: botany, production and uses. CABI Publishing, Wallingford, UK, pp. 253-274.

Bertolini, R., 2004. Strategic thinking: making information and communication technologies work for food security in Africa. Available at: http://www.ifpri.org/pubs/ib/ib27.pdf.

Brown, B.I., 1986. Temperature management and chilling injury of tropical and subtropical fruit. Physiology of the tree. Acta Horticulturae 175: 339-342.

De Carvalho, L.M.J, De Castro, I.M. and Da Silva, C.A.B., 2008. A study of retention of sugars in the process of clarification of pineapple juice (*Ananas comosus*, L. Merril) by micro- and ultra-filtration. Food Engineering 87: 447-454.

Collins, J.L., 1968. The pineapple, botany, utilization, cultivation. Leonard Hill Ltd, London, UK, pp. 200-294.

Daniel, J. and Martin, T., 2008. Impacts des normes obligatoires de la réglementation européenne, d'agriculture biologique et du commerce équitable sur les systèmes de production d'ananas au Bénin et au Togo. Rapport de mission Normes, CE-MFAE, 83 pp.

Dohou, V.B., 2008. Programme national de développement de la filière ananas. ADEX Rapport final, pp. 15-37. Available at: http://tinyurl.com/jzmudye.

Dolan, C. and Humphrey, J., 2000. Governance and trade in fresh vegetables: the impact of UK supermarkets on the African horticulture industry. Journal of Development Studies 37: 147-176.

Dorward, A., Fan, S., Kydd, J., Lofgren, H., Morrison, J., Poulton, C., Rao, N., Smith, L., Tchale, H., Thorat, S., Urey, I. and Wobst, P., 2004. Institutions and policies for pro-poor agricultural growth. Development Policy Review 22: 611-622.

FAOSTAT, 2012. Food and agriculture organization of the United Nations, crop and livestock products trade. Available at: http://tinyurl.com/owkulmk.

Food and Agriculture Organization of the United Nations (FAO), 1997. Guidelines for small-scale fruit and vegetable processors. FAO Agricultural Services Bulletin 127. Available at: http://tinyurl.com/oxh7kbq.

Fassinou, H.V.N., Lommen, W.J.M., Van der Vorst, J.G.A.J, Agbossou, E.K. and Struik, P.C., 2012. Analysis of pineapple production systems in Benin. In: Wünsche, J.-N. and Albrigo, L.G. (eds.) Proceedings of XXVIIIth IHC-IS on citrus, bananas and other tropical fruits under subtropical conditions. Acta Horticulturae 928, ISHS 2012, pp. 47-58.

Garnier, C.L., 1997. Ananas: la 'queen Tahiti', Nnte technique du département de la recherche agronomique appliqué. Tahiti, Polynésie Française, pp. 1-14.

Gbenou, R.K., Taoré, M. and Sissinto, E., 2006. Etude accélérée de marché (EAM) sur les différents produits ananas au Bénin. Working paper, Helvetas, Benin, Cotonou, Benin.

Gibbons, R., 2005. Incentives between firms (and within). Management Science 51: 2-17.

Gray, V., 2006. The un-wired continent: Africa's mobile success story. Available at: http://tinyurl.com/p9xnvls.

Helvetas-Benin, 2007. Appui à la filière ananas biologique et équitable: document du projet. Helvetas-Benin, Cotonou, Benin, pp. 6-9.

Henson, S. and Reardon, T., 2005. Private agri-food standards: implications for food policy and the agri-food system. Food Policy 30: 241-253.

Hepton, A., 2003. Cultural system. In: Bartholomew, D.P., Paull. R.E. and Rohrbach, K.G. (eds.) The pineapple: botany, production and uses. CABI Publishing, Wallingford, UK, pp. 109-142.

Hounhouigan, M., 2014. Quality of pasteurised pineapple juice in the context of the Beninese marketing system. Wageningen University, Wageningen, the Netherlands.

Ingenbleek, P.T., Tessema, W.K., Van Trijp, H.C., 2013. Conducting field research in subsistence markets, with an application to market orientation in the context of Ethiopian pastoralists. International journal of Research in Marketing 30: 83-97.

Institut National de la Statistique et de l'Analyse économique (INSAE), 2009. Available at: http://www.insae-bj.org.

International Trade Centre (ITC), 2013. Gestion des Coopératives d'Ananas, Bénin et Togo, Document Technique, 152 pp.

Jaffee, S. and Henson, S., 2004. Standards and agro-food exports from developing countries: rebalancing the debate. World Bank Policy Research, working paper 3348. World Bank, Washington, DC, USA.

Jan, A. and Masih, E.D., 2012. Development and quality evaluation of pineapple juice blend with carrot and orange juice. International Journal of Science and Research Publications 2: 2250-3153.

Linford, M.B., King, N. and Magistad, O.C., 1934. Planting and fruit quality. I. Comparison of large and small slips in pure and mixed stands. Pineapple Quality 4: 176-190.

Malézieux, E., Zhang, J., Sinclair, E. and Bartholomew, D.P., 1994. Predicting pineapple harvest date in different environments, using a computer simulation model. Agronomy Journal 86: 609-617.

Mcmillan, J. and Woodruff, C., 2000. Private order under dysfunctional public order. Michigan Law Review 98: 2421-2458.

Miller, E.V., 1951. Physiological studies of the pineapple, *Ananas comosus. L. Merr.* with special reference to physiological breakdown. Plant Physiology 26: 66-75.

Mongbo, R. and Floquet, A., 2006. Pineapple against poverty? Market opportunities, technological development and social stratification in Southern Benin. October 11-17, 2006. University of Bonn, Tropentag, Germany.

Morton, J.F., 1987. Fruits of warm climates. NYBG, New York, NY, USA.

Ngoddy, P.O. and Ihekoronye, I.A., 1985. Integrated food science and technology for the tropics. Macmillan Publishers, London, UK, pp. 73-303.

Pay, E,. 2009. The market for organic and fair-trade mangoes. Study prepared in the framework of FAO project GCP/RAF/404/GER, 'Increasing incomes and food security of small farmers in West and Central Africa through exports of organic and fair-trade tropical products'. Food and Agriculture Organization of the United Nations, Rome, Italy.

Reardon, T. and Farina, E., 2002. The rise of private food quality and safety standards: illustrations from Brazil. International Food and Agribusiness Management Review 4: 413-421.

Royer, A. and Bijman, J., 2012. Towards an analytical framework linking institutions and quality: evidence from the Beninese pineapple sector. African Journal of Agricultural Research 7: 5344-5356.

Ruben, R., Tilburg, A., Trienekens, J. and Boekel, M., 2007. Linking market integration, supply chain governance, quality and value added in tropical food chains. In: Ruben, R., Van Boekel, M.A.J.S., Van Tilburg, A. and Trienekens, J. (eds.) Tropical food chains: governance regimes for quality management. Wageningen Academic Publishers, Wageningen, the Netherlands, pp. 13-45.

Trienekens, J. and Zuurbier, P., 2008. Quality and safety standards in the food industry, developments and challenges. International Journal of Production Economics 113: 107-122.

Van Boekel, M.A.J.S., 2009. Innovation as science. In: Moskowitz, H.R., Saguy, I.S. and Straus, T. (eds.) An integrated approach to new food product development. CRC Press/Taylor and Francis Group, New York, NY, USA, pp. 37-50.

Wardy, W., Saalia, F.K., Steiner-Asiedu, M., Budu, A.S. and Sefa-Dedeh, S., 2009. A comparison of some physical, chemical and sensory attributes of three pineapple (*Ananas comosus*) varieties grown in Ghana. African Journal of Food Science 3: 22-25.

Williamson, O.E., 1979. Transaction-cost economics: the governance of contractual relations. The journal of Law and Economics 22: 1-233.

Williamson, O.E., 1991. Strategizing, economizing, and economic-organization. Strategic Management Journal 12: 75-94.

Ziggers, G.W. and Trienekens, J., 1999. Quality assurance in food and agribusiness supply chains: developing successful partnerships. International Journal of Production Economics 60: 271-279.

Appendix 3.1. Global SWOT analysis of pineapple value chains (Adossou, 2012; Arinloye, 2013).

Strengths	Weaknesses
• Pineapple yield is the highest in Africa • Producers have experience in pineapple cultivation (at least 20 years) • Presence of (few) companies with good experiences in dried pineapple, pasteurised juice and fresh juice production • Improved pineapple farm-gate price from 85 to 110 CFA Francs in the last 5 years on international (European) markets • Beginning of Société Nationale pour la Promotion Agricole (SONAPRA) involvement in supplying fertilisers (non-specific) through Micro Finance Institutions	• Difficult access to specific inputs such as fertilisers • Difficult access to appropriate equipment, and logistics (cooling chain, packaging, transportation conditions, and labelling) • Lack of technical knowledge of producers • Lack of organisation of the producers • High air freighting cost • High competition and low collective action among exporters
Opportunities	**Threats**
• Growing interest of government in promoting pineapple value chains as diversification strategies • Existence of unvalourised neighbouring countries market: i.e. Nigerian market with currently no formal trader organisation and arrangement • Pineapple production expending to other agro-ecological zone (Oueme-Plateau and Mono-Couffo departments) • New international market opportunities with Kosovo • Beginning of mobile phone (SMS) use market information, thereby reducing asymmetric reduction (underway with vegetables and to be generalised to other value chains including pineapple) in collaboration with ESOKO, Partners for Development (PfD) and EU	• High land pressure especially in south Benin • Importers impose their will within the sector or they are dishonest • The rule of free products and citizens circulation in the Economic Community of West African States (ECOWAS) countries is mostly not respected at the borders • Weak water management and high rainfall dependency • Insufficiency/inexistence of the seedling (suckers and nurseries) especially for 'Smooth cayenne' cultivar 'Cayenne lisse' • Tendency to the extinction of 'Cayenne lisse' cultivar because of the low regeneration rate

Appendix 3.2. Constraints along the pineapple value chain and recommended solutions

	Constraints expressed by farmers	Recommended solutions by farmers
Input supply	• Non-availability (quantity and quality) of vegetal material especially for 'Smooth cayenne' • Lack of fertilisers and pesticides suppliers • Non-availability and high cost of specific fertiliser	• Develop planting materials nurseries and establishing contracts with nursery producers • Establishing contractual agreements between input suppliers, exporters; banks and producers
Production	• High sensibility of 'Smooth cayenne' cultivar to diseases mainly wilt • Non-respect of Ethephon/Ethrel (plant growth regulator) residue by producers • High heterogeneity in pineapple quality • Instability/fluctuation of produced quantity across the year (low January, February and March period because of dry season)	• Train producers on technical itineraries with regular field monitoring • Train producers on request practices to get pineapple with less residues and homogeneous products • Develop agronomic tool (new cultivar) allowing the resistance of vegetal materials to drought and climate change
Conditioning and logistic	• Difficulties of supply boxes (need to travel and face borders constraints to Ghana) • Lack of appropriate logistic equipment (e.g. storage facility in the airport, boat with cooling facilities, cooling chain) • Poor rural roads condition leading to high fruits losses and high transportation cost. This also leads to delay in transport to the airport	• Bulk order of boxes to reduce transaction costs • Technical, infrastructural and financial supports • Need for state involvement in establishing appropriation packaging factories
Processing	• Artisanal mechanic equipment • Need for improving pineapple juice packages • Inconsistent supply and high cost of imported cane packs • Variability in produced juice due to the high heterogeneity in fruit quality produced in the field	• Improve processing practices with modern and hygienic equipment • Investing in cane pack procurement • Laboratory investigation on pasteurisation techniques
Quality issues	• Not good understanding of pineapple attributes by farmer (literacy problem) • Not qualified teams for conditioning activities • Traceability system not well defined and not well understood by stakeholders • Quality control not operational	• Train farmer in quality norms and standards for traceability in local language • Laboratory accreditation on international market norms and standards • Provide laboratories (Direction de l'Alimentation et de la Nutrition Direction de l'Alimentation et de la Nutrition Appliquée (DANA) and Product Development and Quality Control (PDQC) with adequate equipment, knowledge for capacity building

(adapted from Arinloye, 2013)

4. Willingness to pay for market information received by mobile phone among smallholder pineapple farmers in Benin[1]

D.D.A.A. Arinloye[1]*, A.R. Linnemann[2], G. Hagelaar[3], S.W.F. Omta[3], O.N. Coulibaly[4] and M.A.J.S. van Boekel[2]

[1]*World Agroforestry Centre (ICRAF), Bamako Mali, BPE5118 Bamako, Mali;* [2]*Wageningen University, Product Design and Quality Management, Postbus 17, 6700 AA Wageningen, the Netherlands;* [3]*Wageningen University, Management Studies Group, Hollandseweg 1, 6706 KN Wageningen, the Netherlands;* [4]*International Institute of Tropical Agriculture (IITA), Benin Station, 08BP0932 Trip Postal, Cotonou, Benin; a.arinloye@cgiar.org*

Abstract

Access to up-to-date information on market prices and quality requirements remains a key issue for smallholder farmers' access to high income markets. The aim of this chapter is to explore the problem of information asymmetry between farmers and buyers in the pineapple supply chain in Benin, and to assess strategies using mobile phones to overcome this problem. Data was collected from an exploratory case study in Ghana and a survey with 285 farmers in Benin. Results show that farmers face market information asymmetry leading to lower prices and income. In Ghana, market price alerts through mobile phones messaging allowed decreasing transaction costs for farmers. In Benin, farmers expressed a willingness to pay a premium of up to US$ 2.5 per month to get market price and quality information. Econometric analysis showed that decisive factors for the size of the premium include farm location, market channel, profit margin, contact with agricultural extension services, and technical support from buyers.

Keywords: information asymmetry, contingent valuation, food quality, market price, willingness to pay

4.1 Introduction

Recent trends towards higher food safety standards and stricter traceability requirements in key importing countries of agricultural products increase the information asymmetry between buyers and producers, thereby raising the bar for smallholders to enter such markets due to high compliance costs (Suzuki *et al.*, 2011).

---

[1] Parts of this book chapter have been published before under: Arinloye, D.D.A.A., Linnemann, A.R., Hagelaar, G., Coulibaly, O. and Omta, S.W.F., 2015. Taking profit from the growing use of mobile phone in Benin: a contingent valuation approach for market and quality information access. Information Technology for Development 21: 44-66.

Information asymmetry refers to the fact that many transactions are characterised by incomplete, imperfect or unbalanced information among the transactional parties (Claro *et al.*, 2004; Williamson, 1985). The quality of agricultural products and their safety attributes depend on how they were grown in the field; for instance by organic or conventional farming methods. Such information is obviously known to the farmer, but not to third parties, because the cultivation practices cannot be determined simply by looking at the final product (Mikami and Tanaka, 2008). In contrast, buyers in the markets are much better informed about market prices and their fluctuations.

This issue of information asymmetry becomes more important when the number of intermediaries (collectors, middlemen, wholesalers, and retailers) along the supply chain increases. If price information is distributed asymmetrically between farmers and buyers, the market for agricultural products may fail to achieve an efficient resource allocation because of (the risk of) moral hazard or adverse selection (Akerlof, 1970; Holmstrom, 1979; Ozer and Wei, 2006; Resende-Filho and Hurley, 2012). These informational problems could be avoided if farmers had access to accurate market information, like current prices (Mikami, 2007). Reduced information asymmetry between farmers and buyers implies a more informed trade, which, in turn, increases the market impact of the buyers' trades. Hence, farmers may be able to increase their profit by sharing cost information with buyers. When there is high information asymmetry between farmers and buyers, this generally results in low profits for the farmers (Mendelson and Tunca, 2007).

The introduction of mobile phones has brought new possibilities for people to communicate and share information, for instance, on markets and services. The impact of this development was felt across all sub-Saharan African (SSA) countries. For example, in Ghana, farmers in Tamale are able to send a text message to learn about maize, pineapple and tomato prices in Accra, over 433 kilometres away. In Niger, day labourers are able to call acquaintances in Benin to find out about job opportunities without making the US$ 40 trip (Aker and Mbiti, 2010). In Kenya, those affected by HIV and AIDS can receive text messages daily, reminding them to take their medicines on time (Pop-Eleches *et al.*, 2011). Citizens in countries as diverse as Kenya, Nigeria and Mozambique are able to report violent confrontations via text messages to a centralised server that is viewable, in real time, by the entire world (Aker and Mbiti, 2010). In Benin, data has been collected for decades on market prices of food products (ONASA, 2011), but this information fails to reach users (farmers, traders and processors) at the right time in an accessible and usable manner. On the one hand, research suggests interesting strategies on how to promote agricultural extension services to farmers, but on the other hand it is unclear how this information must be managed. Therefore, it was investigated if and how the mobile phone can be used in the pineapple chain, which is one of the promising export crops in Benin, where the above mentioned challenges of information asymmetry are present and where there is limited access to high income markets by smallholders farmers because of the quality norms and standards barriers they face.

Although the increased market information flow (especially on commodity prices) can potentially benefit marketing of all kinds of crops, it has a larger impact on reducing information asymmetry on market prices for perishable products, where quality is strongly related to the freshness at the time of exchange (Kalyebara *et al.*, 2007; Muto and Yamano, 2009). The new flow of information made available by mobile phones in African countries can help farmers and traders by providing accurate market information, allowing them to transport and trade their perishable products quickly to avoid spoilage. Access to information by mobile phones can also help farmers to decide whether or not to accept the price offered by traders, by obtaining price information from other sources.

Mobile phone services that provide accurate and up-to-date market information can be financially supported by governments, development projects, investment programs, or international partners for development (Donner, 2009; Donner and Escobari, 2010; Kizito, 2011). In Mali, for instance, contracting for the provision of market information is at the national level, but with a mix of funding sources from public and private sectors (Kizito, 2011). In most of the cases, however, these services do not sustain after the development and investment programs terminate, which raises doubts as to the (perceived) benefits of these services to users and users' respective willingness to pay for the services.

A number of questions have, so far, not been answered in the literature. First, to which extent are smallholder farmers able and willing to pay a premium to access market information services – excluding external subsidies? Second, does the market price and quality information asymmetry really matter in supply chains? Answering these questions will help design a short message service (SMS) based framework for sustainable and efficient market information systems (MISs) that are easily accessible for smallholder farmers in less developed countries.

With respect to these issues, much has been written on the role of information and communication technologies in Africa with a special focus on factors that affect the spread of mobile coverage and the impact of mobile phone use on pro-poor labour market access, employment creation and health care (Aker, 2008; Bosch, 2009; Brouwer and Brito, 2012; Buys *et al.*, 2009; Lawson-Body *et al.*, 2011; Maranto and Phang, 2010; Porter, 2012; Porter *et al.*, 2012). However, most of these studies did not investigate the perceptions of the subscribers and the premiums they are willing to pay for a sustainable use of the mobile phone as a device to access market information in rural and peri-urban areas.

As elucidated by Donner (2008) and Aker and Mbiti (2010), economic research on smallholders' adoption and use of mobile phones in less developed countries has been limited. Using a contingent valuation approach, the present study aims to assess farmers' willingness to use a mobile phone to supply and receive market and quality information on agricultural products, as well as to investigate the premium that they are willing to pay for these services.

First, an explorative case study was undertaken in Ghana – a country with many years of experience in mobile phone-based market information management – to gain further insight into smallholders' perceptions of SMS-based MIS. Lessons learnt from Ghana were used to design a survey to investigate the premium that pineapple farmers in Benin are willing to pay for receiving SMS-based product price information (hereafter called price-SMS) and SMS-based product quality information, such as information on standards, inputs and crop diseases (hereafter called quality-SMS). The outcome of this study was used to formulate policy and development recommendations for improving market access of smallholder pineapple farmers and agrifood chain actors.

The remainder of the chapter is organised as follows. First, we present an overview of the pineapple supply chain in Benin. Second, we present the analytical framework, and explain the methods we used for data collection and analysis. Third, the major findings and lessons learnt in Ghana and the major findings of the econometric analysis of farmers' willingness to pay for price-SMS and quality-SMS in Benin are presented. The implementation strategies and implications for policy and practitioners are put forward in the last section.

4.2 Overview of pineapple supply chains in Benin

There are five main supply chains for pineapple in Benin: the domestic fresh chain, the domestic juice chain, the regional fresh chain, the international fresh chain and the dried pineapple chain (Figure 4.1).

- *Domestic fresh consumption.* This marketing channel is one of the major outlets in Benin, absorbing about 35% of production in 2010. The produce is sold at urban (Dantokpa in Cotonou) as well as rural (Glo-Djigbé, Sekou, Sèhouè, Zinvié, Ouegbo, Ze) markets.
- *The West African (regional) markets for fresh pineapple.* The supply of pineapples in Benin exceeds national demand. Therefore, producers need to find other marketing channels to sell their surplus pineapples. Although there are no official statistics on the quantity of pineapples exported to neighbourhood countries, it is estimated to be around 40% of the national production. Wholesalers in Dantokpa market (Cotonou) stated that the Nigerian market alone absorbs more than 40% of the national production. Regional export markets operate differently from the European and Asian export markets and are dominated by informal transactions with lower quality requirements.
- *Juice from fresh pineapple.* Pineapple juice is produced in traditional and semi-industrial processing factories and packed in 0.25 or 0.33 litre bottles. This market channel is not well developed and is dominated by individual traditional producers and some farmers' associations. The juice is mainly sold on the domestic market and not exported to Europe, because of shelf-life difficulties. This channel consumes almost 15% of national production (which increases the domestic consumption of fresh pineapples to about 50%).

- *Fresh pineapple exports.* The international market (beyond West Africa) accounts for about 2% of total production. This market includes EU countries (France, Belgium, Luxembourg, Italy, Germany, the Netherlands), Asian countries (United Arab Emirates and Saudi Arabia) and North African countries (Algeria and Libya). Exports, either by air or by sea, are problematic. Until 2008, plane freight cost 518 €/tonne by KLM/Air France and 609 €/tonne on DHL[2]. By boat, the freight cost is 380 €/tonne, less expensive than by plane, but it is necessary to ship quite large batches. International exports require a wide range of additional inputs (boxes, bags, and other packaging materials) to ensure that the perishable fruit is effectively preserved. These inputs need to be available and affordable.
- *Dried pineapple and marmalade export chain.* This market channel is not well developed. The major destinations are France, Switzerland, Belgium and Austria. The Tropical Fruit Drying Centre (CSFT-Benin) is the main factory that supplies dried pineapple export from Benin, including pineapple marmalade and syrup.

Participation in the export chain involves fulfilling certain quality attributes, such as size, sugar content, and the absence of external and internal damage. These attributes determine the price paid. The lowest prices for pineapple in the rural, urban and regional markets are recorded during May and June. One respondent indicated that one of the main causes is market competition with other fruits (oranges, mangoes, and bananas), which ripen in the same period. Farmers selling during this period report prices that are below the costs of production, but these can be compensated for by an increase in price from July to September. During this period, the average price of

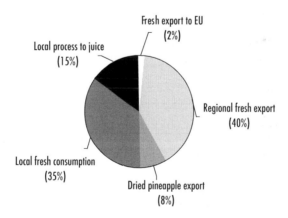

Figure 4.1. Fresh and processed pineapple markets (adapted from Agbo et al., 2008).

---

[2] Data provided by ADEx (*Association des Exportateurs*), the association of exporters in Cotonou-Benin during an interview in 2009.

forty pineapples[3] can be as much as US\$ 30 corresponding to US\$ 0.75/kg (for the 'Smooth cayenne' cultivar).

There are three causes for the annual cycle of price fluctuations. First, pineapple production in Benin mainly depends on natural rainfall patterns that do not allow farmers to apply inputs, mainly fertilisers and ethylene for flowering induction treatment (FIT), during the dry seasons. In south Benin the dry season occurs in December, January and early February. It is difficult to apply the FIT at this time, meaning that there is a shortage of pineapples eight months later, between July and September. Second, there is a socio-cultural condition that affects the profitability of pineapple chains: Muslims' fasting period generally falls in the period between July and September. During this period there is a peak in the demand for fruits, and the local prices experience a significant increase. In the normal season a bunch of forty Cayenne smooth pineapples might fetch between 2,500 and 3,500 CFA Francs[4], or even as little as 1,500 CFA Francs. In the period of Ramadan the same fruit might sell for between 4,000 and 5,000 CFA Francs and large size fruits might even reach 11,000 CFA Francs. The market price for the 'Sugarloaf' variety is normally between 1,500 and 2,000 CFA Francs (for forty), but can increase to 2,500 or 4,000 CFA Francs during the fasting period. Aware of this price fluctuation, farmers now try to manage their production systems so that they can produce during the peak price season. Third, a similar pattern of seasonal demand from other neighbouring countries with significant Muslim populations, such as Nigeria, Burkina Faso and Niger, adds to this high price.

4.3 Data collection and methods

4.3.1 Research models

The analytical framework used in the present study is built around three complementary methods of econometric modelling. We first estimated the determinants of mobile phone use (using a Probit model to take selection bias into account). In the second stage, the factors relevant to explaining farmers' willingness to pay (WTP) for MIS were assessed, using an Ordered Probit model. Finally, the extent to which farmers are ready to pay an affordable price for this service was estimated using a Censured Tobit model approach. This section presents a detailed explanation of each of these analytical approaches.

In general, the endogeneity issue[5] related to the difficulty of disentangling the effect of using a mobile phone (or not) on the willingness to pay for a MIS is a key determinant

---

[3] Selling pineapples in heaps containing forty single pineapples is a common practice among pineapple retailers in Benin. Heaps are sold either on the road (e.g. in Sékou, Zè, Toffo) or in rural and urban markets.

[4] US\$ 1= 502 CFA Francs during data collection in 2009.

[5] For further details on endogeneity issues with endogenous variables see Greene (2008).

in the analytical framework. Hence, rejecting the null hypothesis by observing the significance of the explanatory variables in the model may not imply any causality in terms of farmers' effective WTP. A third driver – the use of a mobile phone – may also affect the dependent variables, inducing a spurious correlation and a selection bias that may lead to erroneous conclusions. The presence of this bias can be tested for by including a sample selection term in the regression. To take account of a possible sample bias that may be related to the inclusion or not of mobile phone users in the model, we first ran a Probit model to generate the Inverse Mill's Ratio (IMR) (2008), which was later on included in the Ordered Probit and Tobit models.

For the Probit model, we define the dependent variable as a dummy with a value of 1 if the farmer has an operating mobile phone and 0 if not. Following White (2004) this leads to a 'selection equation' presented as follows:

$$Z_{ij}^* = \gamma_{0i} + \gamma_{ij}\Sigma W_{ij} + \mu_j \tag{1}$$

where Z_{ij}^* is a variable defining whether the farmer has already access to (and uses) a mobile phone or not, and W_{ij} presents a set of explanatory variables. The IMR is then generated from the parameter estimates of the Probit regression of Equation 1. In the second step, using only the observations of farmers who have and use this technology, and including the IMR as a dependent variable, we estimated the WTP Ordered Probit (Equation 4) and the Tobit (Equation 5) models. For the WTP Ordered Probit, the general analytical framework consists of the following equation:

$$Y_{ij} = \alpha_{0i} + \Sigma\alpha_{ij}X_{ij} + \varepsilon_j \tag{2}$$

where Y_{ij} is the target dependent variable (with 5 level Likert scale responses), X_{ij} is a set of control and independent variables and ε_j is a vector of error terms. More specifically, the null hypothesis is that all the slope coefficients of the explanatory factors (X_{ij}) are equal to zero (H_0: $\alpha_{ij}= 0$). The basic assumption is that a farmer will only express a WTP if he has an operational mobile phone. While $Y_{ij} = s$ (with s = 1-5) implies that the equation has been precisely measured, there exists an unobservable (latent) variable Y_{ij}^* such that $\eta_{ss} \leq Y_{ij} \leq \eta_s$ with s = 1-5. Following Verbeke and Ward (2006), farmers' WTP for the mobile phone-based MIS is expressed as follows:

$$Y_{ij} = \begin{cases} 1 \Rightarrow \text{strongly disagree} & \Rightarrow \text{if } \eta_0 = f \text{ ong} Y_{ij}^* < \eta_1 \\ 2 \text{ disagree} & \Rightarrow \text{if } \eta_1 \leq Y_{ij}^* < \eta_2 \\ 3 \text{ indifferent} & \Rightarrow \text{if } \eta_2 \leq Y_{ij}^* < \eta_3 \\ 4 \text{ agree} & \Rightarrow \text{if } \eta_3 \leq Y_{ij}^* < \eta_4 \\ 5 \text{ strongly agree} & \Rightarrow \text{if } \eta_4 \leq Y_{ij}^* < \eta_5 \end{cases} \tag{3}$$

The variable Y_{ij} is observed only when Z_{ij}^* is larger than zero (Equation 1). Hence, the expected farmers' WTP, premised upon the possession of a working mobile phone in the Ordered Probit model is expressed as:

$$E(Y_{ij}|Z_{ij}^* > 0) = \alpha_{0i} + \alpha_{ij} \sum X_{ij} + \sigma_{ij} \frac{\emptyset(X_i^P \alpha_i)}{\Phi(X_i^P \alpha_i)} + \varepsilon_j \qquad (4)$$

Where $\emptyset$ is the probability density function of a univariate normal distribution and Φ is the cumulative distribution function. The term $\emptyset(X_i^P \alpha_i) / \Phi(X_i^P \alpha_i)$ is the IMR.

To assess if the WTP for mobile-based MIS was sufficiently high, farmers were asked the amount of money they would be willing to spend to get that service. If they did not express a WTP of any premium, the measure of desire is zero (Paolisso *et al.*, 2001). Following Maddala and Lahiri (2006), the estimated Tobit model is expressed as follows:

$$E(\pi_{ij}|Z_{ij}^* > 0) = \beta_{ij} \sum X_{ij} + \sigma_{ij} \frac{\emptyset(X_i^P \alpha_i)}{\Phi(X_i^P \alpha_i)} + \varepsilon_j \qquad (5)$$

where π_{ij} is the amount of money i that farmer j is ready to pay to get or supply market information using a mobile phone (assuming current possession of an operational mobile phone) $(Z_{ij}^* > 0)$, X_{ij} is the set of explanatory variables that are hypothesised to affect the amount that farmer j is willing to pay, β_{ij} is the parameter to be estimated and ε_j the error terms' vector.

If the IMR has a significant coefficient in both Equation 4 and 5, this means that running the regression models without differentiating between farmers who are using a mobile phone from those who are not – as a basic condition – would have led to selection bias. Before running the econometric models, each variable was checked for normality using a Skewness and Kurtosis tests (D'agostino *et al.*, 1990).

From the literature, several factors (X_{ij}) are hypothesised as affecting farmers' willingness to adopt innovations (Adegbola and Gardebroek, 2007; Adesina and Zinnah, 1993; Adesina *et al.*, 2000; Binam *et al.*, 2004; Feder *et al.*, 1985; Herath and Takeya, 2003; Sall *et al.*, 2000). These factors include socio-economic characteristics, such as age, farming experience and income or profit (Adegbola and Gardebroek, 2007; Adesina and Zinnah, 1993; Arinloye *et al.*, 2010a). The farmers' dynamic capability, i.e. their aptitude to be flexible in response to the market and environment changes, is also a determinant (Clark and Fujimoto, 1991; Wang and Ahmed, 2007; Woiceshyn and Daellenbach, 2005). The awareness level, which is determined by contact frequency with extension agents and support received or membership of an association, has also been found to significantly affect farmers' willingness to change (Adegbola and Gardebroek, 2007). The institutional environment and market context in which farmers are embedded, also determine their decisions about whether or not to adopt a new technology (Adegbola and Gardebroek, 2007; Thangata and

Alavalapati, 2003). Detailed descriptions of these variables as included in the models and the hypothesised coefficient signs are presented in Appendix 4.1.

4.3.2 Data collection

Data used in this study were collected in two phases. First, an exploratory case study (Yin, 1994) was undertaken in Ghana, predominantly to understand Ghanaian experiences in managing market information with smallholder farmers using mobile phone SMS, and to learn how subscribers perceive and appreciate this innovation in the agrifood sector. During this case study, 45 key informants were interviewed using a non-structured protocol and selected on a non-probabilistic basis. Respondents were chosen on the basis of their experiences and knowledge of the pineapple, production, marketing, supply chain organisation, the use of mobile phone in agriculture, and the existing institutional environment. Detailed information on the categories of actors interviewed in Ghana can be found in Table 4.1. Lessons learnt from this case study in Ghana were used to design a survey in Benin on price-SMS and quality-SMS willingness to pay.

In Benin, data were collected with a pre-tested, semi-structured survey questionnaire, which consisted of a combination of closed questions, Likert scales with a 5-point format (Allen and Seaman, 2007; Jamieson, 2004) and open questions. Figure 4.2 shows the mobile phone network of one mobile phone operator (MTN©) in Benin in 2012. It shows that most of the subscribers are located in southern Benin where our study was undertaken. From the literature (Arinloye *et al.*, 2010b, 2012) we learnt that more than 95% of pineapples produced in Benin are from southern Benin, in particular from the Atlantique Department. Respondents from this area were selected using a randomly stratified sampling scheme (StatPac, 2010). The criteria used were

Table 4.1. Categories of actors interviewed in Ghana.

Categories		Number of respondents
Farmer's organisation leader		3
Individual farmers		21
Traders in local markets		2
Exporters		2
Processing companies		2
Support organisations and institutions	Input suppliers	2
	Ministry of agriculture (government)	4
	University & research centre	2
	Quality control services	3
	Non-governmental organisation	2
	Market information system (ESOKO)	2
Total		45

the acreage under pineapple cultivation in 2009 (differentiated into small scale (<1 ha), medium scale (between 1 ha-5 ha) and large scale (>5 ha)), the supplied market channels (local or export markets), the location of the pineapple farm (i.e. distance to the main market centre in Cotonou, see Figure 4.2) and the support of extension agents. Farmers were contacted with the assistance of agricultural extension officers, who provided the names and addresses of lead farmers in the villages where they intervene. Pineapple producers' associations and councils constituted a second source of information on pineapple farmers.

After data collection, incomplete questionnaires and non-qualifying respondents (i.e. farmers who did not provide accurate information) were eliminated, resulting in a final list of 285 observations in Benin. For data analysis we combined both descriptive and econometric approaches.

To design the WTP questions and assess the premium that farmers are willing to pay, we set a maximum affordable amount in order to avoid exaggerated and uncontrolled answers from respondents. The amount that was fixed, was based on a World Bank survey (World Bank, 2010), that estimated the affordable tariff for a prepaid mobile phone to be US$ 8 per month in the sub-region. This served as a reference to fix the maximum premium threshold at 4,000 CFA Franc (US$ 7.96) per month.

A correlation matrix and the descriptive statistics of the variables included in the models are presented in Appendix 4.2. The correlation coefficients were less than 0.4, generally indicating weak relations (Peters *et al.*, 1997). This clearly shows that

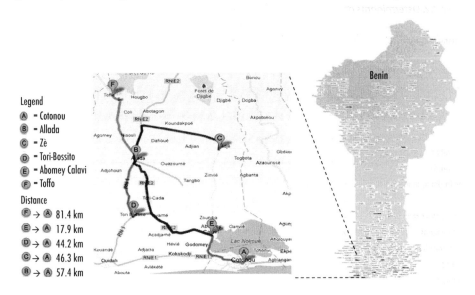

Figure 4.2. Mobile phone network in Benin with study areas, and distance to the main urban market in the south of the country (Adapted from MTN-Benin, 2012).

the variables were sufficiently independent to be modelled without multicollinearity problems (Verbeek, 2008). We used STATA SE software (Statacorp, College Station, TX, USA), which also controlled for the models' robustness – using the *robust* option. The Robust standard errors are reported in Table 4.2.

4.4 Mobile phone-based market information system experiences in Ghana: Esoko case study

The exploratory case study in Ghana was aimed at gaining insights into smallholders' perceptions about an existing SMS-based market information system. Esoko – formerly known as TradeNet – is an agricultural market information platform created in 2006 with the objective to disseminate useful market information to smallholder farmers in less developed countries (https://esoko.com). The organisation is active in 16 East and West African countries, including Ghana. It is a response to the explosive growth of mobile services in Africa. Esoko is a private initiative based in Accra, Ghana, supported by a team of over 60 local developers and support staff. Although the knowledge that farmers have is often underestimated, an asymmetry of information exists throughout agriculture, which rewards some and excludes others. To overcome this situation, Esoko assists smallholder farmers by providing them with a package of weekly advisory services including current market prices, matching bids and offers, weather forecasts, and news and tips.

Table 4.2. Determinants of mobile phone use, willingness to pay (WTP) for SMS on quality and price.[1]

Variables	Determinant of mobile phone use	Determinant of WTP for SMS on quality	Determinant of WTP for SMS on price
Socio-economic and farm characteristics	age	–	age
	education level	–	–
	profit margin	–	profit margin
	farm size	farm size	farm size
Market attributes	distance	distance	distance
Market channels targeted	–	local market	local market
Institutional support	public support	public support	public support
	–	quality support	quality support
Inverse Mill's ratio (IMR)	–	-0.85 (1.81)	-1.59 (1.98)
Observations	285	247	247
Wald chi^2(df)	53.4(16)***	90.51 (17)***	53.46 (17)***
Pseudo R^2	0.38	0.14	0.12
Log pseudolikelihood	-69.14	-212.93	-181.32

[1] Numbers between brackets are robust standard errors; *** significant at 1%.

How does the Esoko platform help Ghanaian farmers? When questioned, farmers answered that the SMS services help them to improve their price negotiation capacities, find alternative markets, and enable them to sell timely at better prices. The platform provides automatic and personalised price alerts, buy and sell offers, bulk SMS messaging and stock counts. Services provided have transformed mobile phones into market bulletins and increased their utility beyond voice and text. It has succeeded mainly because it allows text messages to be sent and received in several languages, including local languages, and provides real-time commodity prices. Mobile phone applications include the provision of market information and electronic trading platforms, where farmers and traders can access information on commodities being (or to be) sold, their prices, the identity of their buyers and extension service messages.

Like all businesses, farming is based on having the right information at the right time. Farmers need to know what crops to plant to obtain the best return on their investment of time and money. Ghanaian farmers have shown interest in using their mobile phones to get a good yield, and in accessing the appropriate fertilisers and pesticides to apply to their crops. SMS-based market information is also helpful for buyers who sometimes have no information about what is growing where and in what quantity. Esoko has been able to respond to this demand by providing accurate and updated prices, offers and profiles. This data can be accessed by any mobile phone user anywhere in the country covered by the mobile phone network. SMS alerts are sent out either as-they-happen (offers to buy and sell) or on specific days of the week (prices), depending on the subscriber's preference. For farmers, text messages by phone were helpful in reducing costs for searching for information and significantly reduced information asymmetry and misunderstandings with their buyers.

However, the major challenge expressed by illiterate farmers (more than 40% do not have formal education and 35% have less than primary school level) was that they always have to ask the assistance of their children or neighbours to help them to read or send messages. In rural areas this is sometimes coupled with a lack of infrastructure, such as electricity to charge phones.

4.5 Information asymmetry and importance of mobile phone use by smallholder pineapple farmers in Benin

As stated in the introduction, market information asymmetry is a major factor affecting farmers' income in agrifood chains. As evidence, the price of pineapple at the farm gate is generally very low compared to the price at which it is sold to consumers, even in the same area. For example, our investigation shows that the price of 40 medium-sized pineapples of the 'Smooth cayenne' variety (i.e. about 50 kg) varies between 2,500 CFA Francs (US$ 5) and 9,000 CFA Francs (US$ 18) at different periods on the local market, and can even reach 10,000 CFA Francs (US$ 20) during the Ramadan, the fasting period of Muslims, when demand is high. Farmers receive on average only 3,500 CFA Francs (US$ 7) of this. Medium and large-sized pineapples of the second variety, 'Sugarloaf', were sold to consumers on local markets for prices between 1,500 CFA Francs (US$ 3) and 4,500 CFA Francs (US$ 9) and this can reach a pick of 15,000 CFA Francs (US$ 30), while the average farm-gate price is 2,200 CFA Francs (US$ 4.3) for about 40 kg. This shows how variable and unstable the market price can be in the same location. Farmers indicated that they were not aware of the prices at which traders resell their products. For instance, in the survey area, more than 86% of farmers do not know the price paid by the third buyer of their products, as traders do not reveal such information to them. The other 14% of farmers who are (indirectly) informed about traders' market prices, either get the information by travelling to these markets to sell other agricultural products, such as maize, cowpea, and cassava, or by calling their relatives on these markets. This information asymmetry issue is reinforced by their low bargaining power in pineapple transactions (Arinloye *et al.*, 2012). The consequence is that farmers do not know what pineapple farmers in other villages were paid.

As witnessed in Ghana, an SMS-based platform that provides farmers with up-to-date market prices and also asks questions and receives answers from a remote computer-based platform could be a solution to these problems. This would allow farmers to have more information and therefore more bargaining power in their transactions with traders. This platform can match farmers' queries with a database of information about prices in local, urban and regional markets and send answers back to the farmers. Critical market information, such as price, offers, inventories, questions and answers about diseases, can be uploaded and shared through SMS by anyone with a mobile phone. The present study in Benin sheds light on farmers' responsiveness to a mobile phone based MIS.

First of all, it is important to know the proportion of smallholder farmers who are currently using a mobile phone as a communication tool in the study area. Our result shows that the use of mobile phone is widespread in the rural areas in Benin as reflected by the sample of pineapple farmers. On average, 87% of the sample use a mobile phone (Appendix 4.1), a value which does not differ much from the subscription proportion (80%) in SSA (World Bank and ITU, 2012). This can be explained by the increasing network coverage in rural areas. As shown in Figure 4.2, the population covered by Benin's five service providers (MTN©, Moov©, BBCom©, Libercom©, and Glo©), in 2010 was estimated to be 90%, much higher than in SSA in 2009 when it was estimated at 53% (World Bank, 2011; World Bank and ITU, 2012). Several factors can explain this high mobile phone adoption rate: falling communication costs (Sey, 2010), population density, increasing per capita income, and, especially, competition among mobile phone operators (as demonstrated by several authors (Aker, 2008; Aker and Mbiti, 2010; Demirhan *et al.*, 2006; Lin *et al.*, 2011).

In general, most pineapple farmers were positive about using their mobile phone to access and supply market information (4.4 on a 5-point scale). In other words, farmers (strongly) agreed about using their mobile for receiving and supplying market prices, and offering their products to potential buyers all over SSA (at least in the countries covered by Esoko). Farmers also expressed a high level of interest (4.3 on a 5-point scale) in using this tool to get information that could help them improving their product quality and meeting market standards, such as information on agricultural practices, input supply, quality control and questions/answers on disease control.

The descriptive statistics show that farmers are generally willing to pay an average premium of 1,268 CFA Francs (US$ 2.5) per month to get price-SMS and almost the same average price (1,200 CFA Francs ~ US$ 2.4) to receive quality-SMS. This shows that farmers are equally interested in both product price and product quality information.

4.6 Farmers' willingness to pay for a mobile-based market information system in Benin

As presented in Table 4.2, the IMR was not significant for the WTP for either the price-SMS, or quality-SMS. This implies that there was no need to consider selection bias issues by including users and non-users of mobile phone in the models. In other words, both current and potential mobile phone users were highly interested in paying to get and supply information via SMS. The Wald test examines whether any of the parameters of the model that currently have non-zero values could be set to zero without any statistically significant loss in the model's overall goodness of fit (α_{1j} = α_{2j} = α_{3j} = jtα_{ij} = 0). It tests the overall significance of the variables included in the econometric models (McGeorge *et al.*, 1997; Ryan and Watson, 2009). Results show that the Wald Chi^2 is statistically significant at the 1% level, which indicates that the set of coefficients of the model are jointly significant and that the explanatory power of the factors included in the model is satisfactory.

4.6.1 Determinants of mobile phone use

The Probit model of the determinants of mobile use shows that farmers' age, education level, profit margin, farm size, distance to the urban centre and contact frequency with public extension service agents, are significantly correlated with the mobile phone usage in Benin. Among these factors, education level, profit, and contact frequency with extension service agents showed a positive correlation with the adoption at a 1% significance level. In other words, farmers who use a mobile phone mostly have a higher education level, higher farming profit margins and more frequent contact with the extension service.

The results also show that mobile phone users are mostly younger, located close to the main roads and urban centres and produce on small-sized farms. These findings are in line with the expected correlation coefficient sign (Appendix 4.1) and add to the existing literature, especially the publications of Aker and Mbiti (2010) and Buys *et al.* (2009), who have found that the mobile network coverage probability is positively related to income per capita, closeness to the main urban centres and to the main road. Most of the mobile phone users are smallholder farmers, which does not come as a surprise since 88% of the farmers produce pineapple on less than 5 hectares (Arinloye *et al.*, 2012).

4.6.2 Determinants of farmers' willingness-to-pay for quality-SMS and price-SMS

The results of the econometric model of the factors that affect farmers' WTP for SMS based-quality showed that farmers who are most likely to pay for these services are smallholder famers who are located far from the urban centre (Cotonou), trade mostly with buyers coming from urban markets, and have little contact with the agricultural extension service (Table 4.2). In most of the cases these farmers have either received technical support for on-farm quality improvement from their buyers or from non-governmental organisations (NGOs). In fact, most farmers selling to exporters and some urban wholesalers have specific contracting farming arrangements with their buyers (the outgrowing scheme, Arinloye *et al.*, 2012), who provide technical or financial assistance in terms of training, input supply and loans to support the outgrowers and help them meet their specific quality requirements. We can therefore conclude that those who are highly interested in quality-SMS, are farmers with past experiences of having received capacity building or training on product quality improvement and who are aware of the importance of product quality in the supply chain.

Apart from the distance to the urban centre, all the factors that affect farmers' WTP for quality-SMS also significantly affect the WTP to pay for price-SMS, with the same coefficient signs. This implies that farmers who are willing to pay for these services are also smallholder farmers, located far from the urban centre, not trading with local market traders but with those coming from urban or regional areas, having little contact with agricultural extension services and receiving technical support for on-farm quality improvement from their buyers. Additionally, they are mostly smallholder farmers with lower farming profit margins ($P<0.05$) than the average pineapple profit in the study area, which is estimated at 400,000 CFA Francs (US$ 795) per cropping campaign.

4.6.3 Premium to be paid for quality-SMS and price-SMS

Since the results from the Probit and Ordered Probit models presented so far do not allow isolating the marginal effects of each explanatory variable associated with the expected premium (amount) to be paid for both services, we ran a Censored Tobit regression. The goal was to determine how much each set of regressors, such as socio-economic characteristics, market attributes, marketing channels and institutional support received, accounts for farmers' WTP.

Here also, the IMR are not significant, implying that there was no need to consider selection bias issues in the Tobit models. Results show that the F statistics are statistically significant at the 1% level indicating that the subsets of coefficients of the model are jointly significant and the explanatory power of the factors included in the model is satisfactory.

The marginal effect of the factors included in both Tobit models and their significance level are presented in Table 4.3. In terms of socio-economic characteristics, an increase in farmers' age by one year would decrease the premium they are ready to pay by 28 CFA Francs (US$ 0.05) per month for quality-SMS and by 36 CFA Francs (US$ 0.07) per month for price-SMS. This confirms the result of the ordered Probit model of WTP, which indicated that younger farmers are more willing to pay a higher price than older and experienced farmers. Apparently they are also inclined to pay a higher price for price-SMS than for quality-SMS. This can be explained by young farmers having a longer planning horizon and being more willing to take risks (Zegeye *et al.*, 2001). Moreover, farmers who showed a dynamic capability (e.g. having changed their farming practices in response to market and environmental changes to meet their buyers' requirements in the last five years) are willing to pay an additional premium of 371 CFA Francs (US$ 0.74) per month for quality-SMS and even more (394 CFA Francs ~ US$ 0.78 per month) for price-SMS than farmers who showed less dynamic capability. As for the farm size, we found that a reduction of the covered land by one hectare led to an increase of the accepted premium of 183 CFA Francs (US$ 0.36) per month for quality-SMS. A reason for this might be the predominance of pineapple supply chain by small-scale farmers mostly cropping less than one hectare of pineapple. Most of these farmers have shown more interest in the use of mobile phone to get price and quality information as they are the most concerned by this lack of information as compared to the very few big farmers. The pineapple farm ratio indicates farmers' cropping diversification (or specialisation). The results showed that an increase of diversity by 1% leads to an increase of the acceptable premium of 867 CFA Francs (US$ 1.73) per month for quality-SMS. This can be explained by the fact those farmers, with diversified production system, think beyond and have seen the application and relevance of this SMS service in other value chains (i.e. maize, cashew, cassava, shea) which are also affected by weak access to market information and demand attributes especially for international markets. The issue of market information asymmetry is not only observed in pineapple chain, but along the agriculture sector in Benin.

When looking at the market attribute factors, an increase of the distance between farm and main market centre by 1 km, decreases the premium that farmers would be willing to pay for price-SMS by 15 CFA Francs (US$ 0.03) per month. As far as the institutional support factors are concerned, farmers having regular contact with extension agents showed an interest in paying a higher premium of 536 CFA Francs (US$ 1.06) per month for quality-SMS and 257 CFA Francs (US$ 0.51) per month for price-SMS compared to those who do not have this contact. Moreover, farmers who have received support for quality improvement of their products would pay an additional premium of 330 CFA Francs (US$ 0.65) per month for quality-SMS and 132 CFA Francs (US$ 0.26) per month for price and offer SMS compared to those without any quality support experience. Summarising, farmers who are most willing

Table 4.3. Marginal effects after Tobit models for expected premium to be paid (in CFA Francs) for quality and price SMS.[1]

Variables		Premium for quality-SMS	Premium for price-SMS
Socio-economic	age	-28.0[***]	-35.8[***]
and firm characteristics	education level	-6.7	-44.8
	farming experience	1.2	18.1
	dynamic capability	370.9[**]	394.2[**]
	profit margin	-183.3[**]	-80.5
	farm size	-183.3[***]	-181.8
	pineapple farm ratio	-867.1[***]	-187.2
Market attributes	information time	22.4	25.5
	distance	-5.1	-14.8[***]
Market channels	export market	43.9	403.9
	local market	12.5	44.1
Institutional support	extension service support	536.1[***]	256.9[*]
	market support	58.2	73.6
	quality support	330.1[***]	131.7[**]
	farming support	-23.7	75.2
	input support	-3.1	25.5
Inverse Mill's Ratio (IMR)		1,523.1	2,747.1
Observations		247	234
F statistic (df1; df2)		4.9 (17; 230)[***]	4.13 (17;217)[***]
Log pseudolikelihood		-1,849.1	-1,730.6

[1] [*] Significant at 10%; [**] significant at 5%; [***] significant at 1%.

to pay for quality and price SMS are small-scale young famers, showing dynamic capability in improving and diversifying their agricultural practices and production systems, and located closest the city centre and urban markets.

4.7 Concluding remarks

The present study assesses the determinants of farmers' willingness to use a mobile phone to supply and receive market information on the price and quality requirements for agricultural products, and the premium they are willing to pay for these services. This would be a useful strategy for overcoming information asymmetry in the pineapple supply chain. Using an exploratory case study in Ghana to gain insights into smallholders' perceptions about SMS-based market information systems, followed by an in-depth survey in Benin, the results showed the high potential of mobile phones to improve smallholder agriculture in rural areas of SSA. In Ghana, and other countries where Esoko is active, such a system allows farmers to get market information at the right time. Lessons learnt from this case study may be of great importance in developing and promoting agrifood quality improvement and market access, not only in Benin but also across other SSA countries that face the same challenges. Despite the existence of national institutes and support services involved in quality control and strengthening actors' capacity to comply with quality standards, there is a clear need to design a better mechanism for coordinating the supply chain. If small scale producers are to respond to the quality norms and standards for regional and international markets they need to make investments in their production. Additional investments, either from state, financial partners or NGOs, are needed for building roads, cold chain facilities, safe handling and storage facilities, chemical waste disposal pits, hand washing facilities, personal protective equipment, knapsack sprayers, and certified planting material. Pineapple production in Benin is recognised as having a huge potential. The supply chain is showing an increased international orientation despite the low adaptive capacities of smallholder farmers to comply with foreign norms and quality standards.

Even when mobile phones can enhance access to resources and information, they cannot replace investments in public goods, such as roads, electricity and water. In the absence of a proper infrastructure, smallholder farmers will face problems with efficiency and competitiveness (Roberts and Grover, 2012). As such, it is unrealistic to rely on improved access to market information as the only strategy for improving chain performance by smallholder farmers. A mobile MIS approach needs to be embedded in an enabling political and institutional environment, involving value chain actors to find a holistic solution to the pending issues of information asymmetric and market access. Poor infrastructure remains an obstacle to the development of many communities. Markets with a surplus are disconnected from markets with a deficit (and vice versa). Over the last twenty years the Beninese government through ONASA and INSAE[6] has been collecting information about markets, but has not created the channels to deliver this information to the general public or to farmers, certainly not at a speed to make it commercially valuable.

Implementing the mobile-based MIS, while simultaneously improving related infrastructures, may significantly contribute to helping rural communities to improve their livelihoods by achieving a better product quality and facilitating market access at national and continental levels. Such recommendations have been made by several authors (Cavatassi *et al.*, 2011; Mwesige, 2010; Thiele *et al.*, 2009, 2011), who call for multi-stakeholder platforms that will strengthen public and private actors' partnerships and enable smallholders to gain sustainable access to high income markets. The private sector could provide platform coordination and management staff (like Esoko is doing), important value chain actors (such as farmers' organisations) and

---

[6] ONASA refers to *Office National d'Appui à la Sécurité Alimentaire* an equivalent of national office of food security support. INSAE is the *Institut National de Statisque et de l'Analyse Economique*, corresponding to the national institute of statistic and economic analysis of Benin.

a mobile phone operator can serve as the intermediary between subscribers and the computer-based platform. The public sector could provide support through existing national statistical and market information management institutes (for monitoring the collection of and profiling market information) and research institutes and quality control services (to provide reliable answers to chain actors' requests on quality, inputs, and diseases). It could also provide support services that monitor and build the capacity of smallholders and the infrastructure facilities that they need – such as rural roads, packaging and cooling facilities, and finance. As suggested by White (2004), this would create an enabling environment for innovation and help deliver the resources required to build a complex multidimensional and dynamic range of knowledge, skills, actors, institutions and policy within specific political-policy structures capable of transforming knowledge into useful processes, products and services for agriculture. These recommendations could serve as a guideline for policy-makers and practitioners.

Even though farmers showed a high willingness to pay for a mobile phone-based MIS, it remains important to assess how the existing infrastructure and institutional environment can support such a process and make it effective. This offers opportunities for future development and policy-oriented research. One important limitation of the present study is that farmers' dynamic capabilities have been measured by asking them if they have ever changed their farming practices. Having changed farming practices, possibly only once and only slightly, does not necessarily show the dynamic attitude of the farmers. Entrepreneurship attitude could be measured by asking about farmers' changes in market orientation, in realising new resource configurations, strategies and organisation routines (Eisenhardt and Martin, 2000; Yin *et al.*, 2013), as well as about their flexibility in addressing rapidly changing environments (Teece *et al.*, 1994). Future investigations could put some emphasis on these aspects.

References

Adegbola, P. and Gardebroek, C., 2007. The effect of information sources on technology adoption and modification decisions. Agricultural Economics 37: 55-65.

Adesina, A.A. and Zinnah, M.M., 1993. Technology characteristics, farmers' perceptions and adoption decisions: a tobit model application in Sierra Leone. Agricultural Economics 9: 297-311.

Adesina, A.A., Mbila, D., Nkamleu, G.B. and Endamana, D., 2000. Econometric analysis of the determinants of adoption of alley farming by farmers in the forest zone of southwest Cameroon. Agriculture, Ecosystems and Environment 80: 255-265.

Agbo, B., Agbota, G.E.S. and Akele, O., 2008. Atelier de validation de la stratégie et de l'évaluation de plan d'action de la filière ananas au Bénin. Working paper. Cotonou, Benin.

Aker, J., 2008. Does digital divide or provide? The impact of cell phones on grain markets in Niger. BREAD Working Paper No 177. BREAD Washington, DC, USA.

Aker, J. and Mbiti, I., 2010. Mobile phones and economic development in Africa. Journal of Economic Perspectives 24: 207-232.

Akerlof, G.A., 1970. The market for 'lemons': quality uncertainty and the market mechanism. Quarterly Journal of Economics 84: 488-500.

Allen, I.E. and Seaman, C.A., 2007. Likert scales and data analyses. Quality Progress 40: 64-65.

Arinloye, D.D.A.A., Adegbola, P.Y., Biaou, G. and Coulibaly, O., 2010a. Evaluation des strategies optimales de production et de transformation du riz a travers une analyse du consentement des consommateurs au Benin: application du model hedonique. In: Joint 3rd African Association of Agricultural Economists (AAAE) and 48th Agricultural Economists Association of South Africa (AEASA) conference. Cape Town, South Africa, pp. 1-23.

Arinloye, D.D.A.A., Hagelaar, G., Linnemann, A.R., Royer, A., Coulibaly, O., Omta, S.W.F. and Van Boekel, M.A.J.S., 2010b. Constraints and challenges in governance and channel choice in tropical agri-food chains: evidences from pineapple supply chains in Benin. In: 2nd Agricultural science week of West and Central Africa (CORAF/WECARD). Cotonou, Benin, pp. 1-21.

Arinloye, D.D.A.A., Hagelaar, G., Linnemann, A.R., Pascucci, S., Coulibaly, O., Omta, S.F.W. and Van Boekel, M.A.J.S., 2012. Multi-governance choices by smallholder farmers in the pineapple supply chain in Benin: an application of transaction cost theory. African Journal of Business Management 6: 10320-10331.

Binam, J.N., Tonyè, J., Nyambi, G. and Akoa, M., 2004. Factors affecting the technical efficiency among smallholder farmers in the slash and burn agriculture zone of Cameroon. Food Policy 29: 531-545.

Bosch, T.E., 2009. Cell phones for health in South Africa. In: Wels, H. (ed.) Health communication in Southern Africa: engaging with social and cultural diversity. SAVUSA/Rozenberg/UNISA Press Series, Amsterdam, the Netherlands.

Brouwer, R. and Brito, L., 2012. Cellular phones in Mozambique: who has them and who doesn't? Technological Forecasting and Social Change 79: 231-243.

Buys, P., Dasgupta, S., Thomas, T.S. and Wheeler, D., 2009. Determinants of a digital divide in Sub-Saharan Africa: a spatial econometric analysis of cell phone coverage. World Development 37: 1494-1505.

Cavatassi, R., Gonzàlez-Flores, M., Winters, P., Andrade-Piedra, J., Espinosa, P. and Thiele, G., 2011. Linking smallholders to the new agricultural economy: The case of the plataformas de concertación in Ecuador. Journal of Development Studies 47: 1545-1573.

Clark, K.B. and Fujimoto, T., 1991. Product development performance: strategy, organization, and management in the world auto industry. Harvard Business Press, Boston, MA, USA.

Claro, D.P., Zylbersztajn, D. and Omta, S.W.F., 2004. How to manage a long-term buyer-supplier relationship successfully? The impact of network information on long-term buyer-supplier relationships in the Dutch potted plant and flower industry. Journal on Chain and Network Science 4: 7-24.

D'agostino, R.B., Belanger, A. and D'agostino Jr, R.B., 1990. A suggestion for using powerful and informative tests of normality. American Statistician 44: 316-321.

Demirhan, D., Jacob, V.S. and Raghunathan, S., 2006. Information technology investment strategies under declining technology cost. Journal of Management Information Systems 22: 321-350.

Donner, J., 2008. Research approaches to mobile use in the developing world: a review of the literature. The Information Society 24: 140-159.

Donner, J., 2009. Mobile-based livelihood services in Africa: pilots and early deployments. Communication technologies in Latin America and Africa: a multidisciplinary perspective, pp. 37-58.

Donner, J. and Escobari, M.X., 2010. A review of evidence on mobile use by micro and small enterprises in developing countries. Journal of International Development 22: 641-658.

Eisenhardt, K.M. and Martin, J.A., 2000. Dynamic capabilities: what are they? Strategic Management Journal 21: 1105-1121.

Feder, G., Just, R.E. and Zilberman, D., 1985. Adoption of agricultural innovations in developing countries: a survey. Economic Development and Cultural Change 33: 255-298.

Greene, W.H., 2008. Econometric analysis. Pearson/Prentice Hall, Upper Saddle River, NJ, USA.

Herath, P. and Takeya, H., 2003. Factors determining intercropping by rubber smallholders in Sri Lanka: a logit analysis. Agricultural Economics 29: 159-168.

Holmstrom, B., 1979. Moral hazard and observability. The Bell Journal of Economics: 74-91.

Jamieson, S., 2004. Likert scales: how to (ab) use them. Medical Education 38: 1217-1218.

Kalyebara, R., Nkub, J.M., Byabachwezi, M.S.R., Kikulwe, E.M. and Edmeades, S., 2007. Overview of banana economy in the lake Victoria regions of Uganda. In: Smale, M. and Tushemereirwe, W. (eds.) An economic assessment of banana genetic improvement and innovation in the lake Victoria region of Uganda and Tanzania. International Food Policy Research Institute (IFPRI), Washington, DC, USA.

Kizito, A.M., 2011. The structure, conduct, and performance of agricultural market information systems in Sub-Saharan Africa. Michigan State University, East Lansing, MI, USA.

Lawson-Body, A., Willoughby, L., Keengwe, J. and Mukankusi, L., 2011. Cultural factors influencing cell phone adoption in developing countries: a qualitative study. Issues in Information Systems 12: 97-105.

Lin, M., Li, S. and Whinston, A., 2011. Innovation and price competition in a two-sided market. Journal of Management Information Systems 28: 171-202.

Maddala, G.S. and Lahiri, K., 2006. Introduction to econometrics. Wiley, New York, NY, USA.

Maranto, T. and Phang, S., 2010. Mobile phone access in Sub-Saharan Africa: research and a library proposal. Seminar in International and Comparative Librarianship, pp. 1-23.

Mcgeorge, P., Crawford, J.R. and Kelly, S.W., 1997. The relationships between psychometric intelligence and learning in an explicit and an implicit task. Journal of Experimental Psychology: Learning, Memory, and Cognition 23: 239-245.

Mendelson, H. and Tunca, T.I., 2007. Strategic spot trading in supply chains. Management Science 53: 742-759.

Mikami, K., 2007. Asymmetric information and the form of enterprise: capitalist firms and consumer cooperatives. Journal of Institutional and Theoretical Economics 163: 297-312.

Mikami, K. and Tanaka, S., 2008. Food processing business and agriculture cooperatives in Japan: market power and asymmetric information. Asian Economic Journal 22: 83-107.

MTN-Benin, 2012. Couvernature Nationale. Cotonou, Benin. Available at: http://www.mtn.bj/index.php?item=240.

Muto, M. and Yamano, T., 2009. The impact of mobile phone coverage expansion on market participation: panel data evidence from Uganda. World Development 37: 1887-1896.

Mwesige, D., 2010. Working with value chains, using multi-stakeholder processes for capacity development in an agricultural value chain in Uganda. In: Ubels, J., Baddoo, N.A. and Fowle, A. (eds.) Capacity development in practice. Earthscan, Washington, DC, USA.

Office National d'Appui à la Sécurité Alimentaire (ONASA), 2011. Benin-système d'information sur les marchés. Bulletin Mensuel 26. Available at: http://tinyurl.com/p4erpo2.

Ozer, O. and Wei, W., 2006. Strategic commitments for an optimal capacity decision under asymmetric forecast information. Management Science 52: 1238-1257.

Paolisso, M.J., Hallman, K., Haddad, L.J. and Regmi, S., 2001. Does cash crop adoption detract from childcare provision? Evidence from rural Nepal. International Food Policy Research Institute, Washington, DC, USA.

Peters, R.G., Covello, V.T. and Mccallum, D.B., 1997. The determinants of trust and credibility in environmental risk communication: an empirical study. Risk Analysis 17: 43-54.

Pop-Eleches, C., Thirumurthy, H., Habyarimana, J.P., Zivin, J.G., Goldstein, M.P., De Walque, D., Mackeen, L., Haberer, J., Kimaiyo, S. and Sidle, J., 2011. Mobile phone technologies improve adherence to antiretroviral treatment in a resource-limited setting: a randomized controlled trial of text message reminders. Aids 25: 825-834.

Porter, G., 2012. Mobile phones, livelihoods and the poor in Sub-Saharan Africa: review and prospect. Geography Compass 6: 241-259.

Porter, G., Hampshire, K., Abane, A., Munthali, A., Robson, E., Mashiri, M. and Tanle, A., 2012. Youth, mobility and mobile phones in Africa: findings from a three-country study. Information Technology for Development 18: 145-162.

Resende-Filho, M.A. and Hurley, T.M., 2012. Information asymmetry and traceability incentives for food safety. International Journal of Production economics 139: 596-603.

Roberts, N. and Grover, V., 2012. Leveraging information technology infrastructure to facilitate a firm's customer agility and competitive activity: an empirical investigation. Journal of Management Information Systems 28: 231-270.

Ryan, M. and Watson, V., 2009. Comparing welfare estimates from payment card contingent valuation and discrete choice experiments. Health Economics 18: 389-401.

Sall, S., Norman, D. and Featherstone, A., 2000. Quantitative assessment of improved rice variety adoption: the farmer's perspective. Agricultural Systems 66: 129-144.

Sey, A., 2010. Managing the cost of mobile communications in Ghana. In: Fernandez-Ardevol, M. and Hijar, A.R. (eds.) Communication technologies in Latin America and Africa: a multidisciplinary perspective. UOC, Barcelona, Spain, pp. 143-166.

Statpac, 2010. Surveys designing tutorial: survey sampling methods. Lyndale, Bloomington, IN, USA.

Suzuki, A., Jarvis, L.S. and Sexton, R.J., 2011. Partial vertical integration, risk shifting, and product rejection in the high-value export supply chain: the Ghana pineapple sector. World Development 39: 1611-1623.

Teece, D.J., Rumelt, R., Dosi, G. and Winter, S., 1994. Understanding corporate coherence: theory and evidence. Journal of Economic Behavior and Organization 23: 1-30.

Thangata, P.H. and Alavalapati, J., 2003. Agroforestry adoption in southern Malawi: the case of mixed intercropping of Gliricidia sepium and maize. Agricultural Systems 78: 57-72.

Thiele, G., Devaux, A., Reinoso, I., Pico, H., Montesdeoca, F., Pumisacho, M., Andrade-Piedra, J., Velasco, C., Flores, P. and Esprella, R., 2011. Multi-stakeholder platforms for linking small farmers to value chains: evidence from the Andes. International Journal of Agricultural Sustainability 9: 423-433.

Thiele, G., Devaux, A., Reinoso, I., Pico, H., Montesdeoca, F., Pumisacho, M., Velasco, C., Flores, P., Esprella, R. and Manrique, K., 2009. Multi-stakeholder platforms for innovation and coordination in market chains. International Potato Center, Lima, Peru, 431 pp.

Verbeek, M., 2008. A guide to modern econometrics. Wiley, New York, NY, USA.

Verbeke, W. and Ward, R.W., 2006. Consumer interest in information cues denoting quality, traceability and origin: an application of ordered probit models to beef labels. Food Quality and Preference 17: 453-467.

Wang, C.L. and Ahmed, P.K., 2007. Dynamic capabilities: a review and research agenda. International Journal of Management Reviews 9: 31-51.

White, H., 2004. Books, buildings, and learning outcomes: an impact evaluation of World Bank support to basic education in Ghana. World Bank Publications, Washington, DC, USA.

Williamson, O.E., 1985. The economic institutions of capitalism: firms, markets, relational contracting. Free press, New York, NY, USA.

Woiceshyn, J. and Daellenbach, U., 2005. Integrative capability and technology adoption: evidence from oil firms. Industrial and Corporate Change 14: 307-342.

World Bank, 2010. The little data book on information and communication technology, Development Data Group. World Bank Publications, Washington, DC, USA.

World Bank, 2011. The little data book on information and communication technology, Development Data Group. World Bank Publications, Washington, DC, USA.

World Bank and ITU, 2012. The little data book on information and communication technology, Development Data Group. World Bank Publications, Washington, DC, USA.

Yin, H., Guo, T., LI, B., Zhu, J. and Sun, X., 2013. the theoretical analysis model and statistical test research on enterprises development: a study of data from 195 large and medium manufacturing enterprises in China. Przeglad Elektrotechniczny 89: 198-206.

Yin, R.K., 1994. Case study research. Design and methods. Sage publishing, Beverly Hills, CA, USA.

Zegeye, T., Tadesse, B., Tesfaye, S., Nigussie, M., Tanner, D. and Twumasi-Afriyie, S., 2001. Determinants of adoption of improved maize technologies in major maize growing regions of Ethiopia. Enhancing the contribution of maize to food security in Ethiopia. In: Proceedings of the Second National Maize Workshop of Ethiopia. Ethiopian Agricultural Research Organization, Addis Ababa, Ethiopia.

Appendix 4.1. Description of variables and hypothesised signs.

Variables	Description	Value	Hypothesis[1]
Dependent variables			
Use mobile phone	Do you have/use a mobile phone?	1 = yes; 0 = no	
WTP price info SMS	Are you willing to pay for sending/receiving marketing information (price, offers) via SMS?	1 = strongly disagree; 2 = disagree; 3 = indifferent; 4 = agree; 5 = strongly agree	
WTP quality info SMS	Are you willing to pay for sending/receiving quality information (standards, input and disease) via SMS?	1 = strongly disagree; 2 = disagree; 3 = indifferent; 4 = agree; 5 = strongly agree	
Premium for quality info	How much are you willing to pay for quality information (standards, input and disease) via SMS?	continue (CFA Franc/month)[2]	
Premium for price info	How much are willing to pay to send/receive pineapple information (price, offers) via SMS?	continue (CFA Franc/month)[2]	
Independent variables			
Socio-economic and farm characteristics			
Age	Farmer's age	continuous	+/-
Education	Education level of farmer	0 = no (in)formal education; 1 = primary school/informal literacy; 2 = middle school, 3 = high school, 4 = university level	+
Experience	Years in pineapple farming	continue	+
Dynamic capability	Have you ever changed your farming practices in response to market or environment changes to satisfy your buyers?	1 = yes; 0 = no	+
Profit margin	What was your pineapple production profit margin for the last cropping campaign ($\times$1000 CFA Franc)[3]?	0 = <0 CFA Francs; 1 = 0-100; 2= 100-500; 3= 500-1000; 4 = 1000-5,000; 5 = >5,000	+
Farm size	Pineapple farm size in hectare	1 = large scale (>5 ha); 2 = medium scale (1-5) ha; 3 = small scale (<1 ha)	+
Pineapple ratio	Proportion of pineapple land over the total covered land size – farm specialisation	continue [0-1]	+/-
Market attributes			+
Info-time	Time spent to get reliable market information	Number of days	+
Distance	Distance from farm to the central urban market	1 = <30 km; 2 = 30-60 km; 3 = >60 km	-
Bargaining power	Bargaining power of the farmer with buyers	1 = low; 2 = medium; 3 = high	-
Market channel			
Local market	Selling pineapple to local markets	1 = local market; 0 = otherwise	-
Export market	Selling pineapple to export markets	1 = export market; 0 = otherwise	+

Variables	Description	Value	Hypothesis[1]
Institutional support			
Extension service support	Contact with public extension agents	1 = yes; 0 =no	+
Market support	Receiving support to access market (selling)	1 = strongly disagree; 2 = disagree; 3 = indifferent; 4 = agree; 5 = strongly agree	+
Quality support	Receiving support for pineapple quality improvement	1 = strongly disagree; 2 = disagree; 3 = indifferent; 4 = agree; 5 = strongly agree	+
Farming support	Receiving support for farming systems improvement	1 = strongly disagree; 2 = disagree; 3 = indifferent; 4 = agree; 5 = strongly agree	+
Input support	Receiving support to access inputs	1 = strongly disagree; 2 = disagree; 3 = indifferent; 4 = agree; 5 = strongly agree	+

[1] Expected correlation with dependent variables.

[2] Price in CFA Francs/month is generated by asking farmers the amount they are willing to pay per SMS times the frequency of sending/receiving SMS in a month. The threshold of total amount per month is fixed during the survey at a maximum of 4,000 CFA Francs (US$ 7.96) following World Bank (2010).

[3] US$ 1= 502 CFA Francs during data collection in 2009.

Appendix 4.2. Correlation matrix and descriptive statistics of variables.

Variable	Unit	Min	Max	Mean	Std. Dev.	V1	V2	V3	V4	V5	V6	V7	V8	V9	V10	V11	V12	V13	V14	V15	V16
Dependent variables																					
Use mobile phone	(0-1)[1]	0	1	0.87	0.34	1															
WTP price info SMS	(1-5)[2]	1	5	4.4	1.22	0.21	1														
WTP quality info SMS	(1-5)[2]	1	5	4.27	1.28	0.46	0.03	1													
Premium for quality info	Number	0	4,000	1,268	1,137																
Premium for price info	Number	0	4,000	1,200	1,109																
Independent variables																					
Age (V1)	Number	2.83	4.29	3.6	0.28	1															
Education (V2)	(0-4)[2]	0	4	1.05	1.04	0.21	1														
Experience (V3)	Year	2	40	9.99	5.08	0.46	0.03	1													
Dynamic capability (V4)	(0-1)	0	1	0.72	0.45	0.33	0.23	0.3	1												
Profit margin (V5)	CFA Franc[3]	0	5	2.29	1.07	0.33	0.28	0.37	0.31	1											
Farm size (V6)	Ha	1	3	2.33	0.7	-0.4	-0.11	-0.26	-0.16	-0.4	1										
Pineapple farm ratio (V7)	Number	0.02	1	0.46	0.27	-0.04	-0.13	0.02	-0.01	0.24	-0.4	1									
Info-time (V8)	Number	1	30	1.66	2.39	0.03	-0.07	0.08	0.12	0	-0.07	-0.07	1								
Distance (V9)	Number	17.9	81.4	49.98	21.1	0.07	0.22	0.05	0.27	0.26	0.02	-0.06	-0.07	1							
Bargaining power (V10)	(0-1)[1]	0	1	0.13	0.33	-0.05	0.01	0.08	0.06	0.13	-0.01	0.09	0.01	0.18	1						
Local market (V11)	(0-1)[1]	0	1	0.27	0.45	0.09	0.05	0.01	0.12	-0.02	0.08	-0.15	-0.04	0.09	0.26	1					
Public support (V12)	(0-1)[1]	0	1	0.3	0.46	0.21	0.06	0.06	0.19	0.23	-0.17	0.02	-0.02	0.32	0.16	0.25	1				
Market support (V13)	(1-5)[2]	1	5	2.37	1.31	-0.25	-0.06	-0.05	-0.07	-0.13	0.2	-0.06	-0.07	0.05	0.03	-0.08	-0.46	1			
Quality support (V14)	(1-5)[2]	1	5	3.6	1.26	0.17	0.08	0.2	0.13	0.34	-0.19	0.07	-0.1	0.16	-0.03	-0.08	0	0.06	1		
Farming support (V15)	(1-5)[2]	1	5	3.96	1.11	0.32	0.16	0.29	0.22	0.38	-0.34	0.05	0.04	0.16	0.12	0.01	0.38	-0.35	0.43	1	
Input support (V16)	(1-5)[2]	1	5	2.4	1.25	-0.22	-0.08	-0.11	-0.19	-0.18	0.09	-0.08	-0.02	-0.14	-0.03	0.01	-0.08	0.34	-0.26	-0.24	1

[1] Dummy variables.
[2] Categorical variable.
[3] US$ 1= 502 CFA Francs during data collection period.

5. Improving seed potato quality in Ethiopia: a value chain perspective

A. Hirpa[1,2]*, M.P.M. Meuwissen[3], W.J.M. Lommen[4], A.G.J.M. Oude Lansink[3], A. Tsegaye[5] and P.C. Struik[4]

[1]Hawassa University College of Agriculture, Hawassa, Ethiopia; [2]International Institute of Tropical Agriculture, P.O. Box 30258, Lilongwe, Malawi; [3]Wageningen University, Business Economics, Hollandseweg 1, 6706 KN Wageningen, the Netherlands; [4]Wageningen University, Centre for Crop Systems Analysis, P.O. Box 430, 6700 AK Wageningen, the Netherlands; [5]Addis Ababa University P.O. Box 1176, Addis Ababa, Ethiopia; a.tufa@cgiar.org

Abstract

In Ethiopia, use of low-quality seed potatoes by the majority of potato growers is associated with underdevelopment of the seed potato value chains. Three seed potato systems are present in Ethiopia: the informal seed system, the alternative seed system and the formal seed system. This chapter analyses the performance of seed potato value chains with respect to their ability to supply quality seed tubers to seed potato systems, by using the chain performance drivers enabling environment, technology, market structure, chain coordination, farm management, and inputs. Information obtained from literature review, secondary data and key informants' interviews were used for the analysis. In the informal seed system, seed potato value chains suffered from a poor enabling environment such as a low quality technical support and lack of a seed quality control system; use of sub-optimal storage and transportation technologies, sub-optimal farm management practices; and little use of inputs. In the alternative seed system, main constraints were the lack of a seed potato quality control system, poor farm management practices, little use of inputs by seed potato growers, and a distorted seed potato market that resulted from involvement of institutional buyers. Chains in the formal seed potato system were characterised by little involvement of the private and public sectors in the production and supply of seed potatoes. Based on the analysis, improvement options for the three seed systems were identified.

Keywords: seed quality, potato, value chain, Ethiopia

5.1 Introduction

In Ethiopia, agriculture is the main source of livelihood for more than 80% of the population. However, Ethiopian agriculture is characterised by smallholdings and low productivity. The average land holding is about 1 ha (IFAD, 2010; Spielman *et al.*, 2010, 2011; USAID, 2011) and the average productivity of crops and livestock is far below its potential. As a result, food security and cash income are major constraints for smallholder farmers in Ethiopia. One way to improve the current food security and cash income situation is to change to growing crops and rearing livestock with higher productivity potential. Potato (*Solanum tuberosum* L.) is a crop that can be used to improve food security and cash income because of its high yielding ability in a short season, presence of suitable agro-ecological zones within the country, the availability of labour for its production on large areas of land, and the availability of a potential market with considerable added value for its produce (FAO, 2008).

In Ethiopia, farmers grow two types of potato varieties: local and improved. The majority of potato growers grow local potato varieties[7] that are low yielding, infested with diseases and pests, and susceptible to most of the indigenous diseases and pests. Improved potato varieties are high yielding, relatively clean, and disease tolerant. However, seed of improved potato varieties is not available to the majority of the farmers. The majority of potato growers use poor quality seed saved from the previous harvest or obtained from the market. The use of poor quality seed by potato growers is associated with the underdevelopment of the entire potato value chain that results in low production and productivity of potato in Ethiopia. To suggest improvement options, knowledge on the performance of the seed potato value chains is important. However, currently there is a lack of knowledge on the performance of seed potato value chains. The objective of this chapter is to analyse the performance of seed potato value chains with respect to their ability to supply quality seed tubers to seed potato farmers. This chapter attempts to summarise knowledge that can be used to improve seed quality for all seed potato value chains. We characterise how chain activities are performed, we evaluate the performance of the chains against the most important drivers, and identify barriers and options for development.

5.2 Conceptual framework

Chain performance can be analysed by identifying its main drivers and assessing their impacts. In this study we adopted the chain performance drivers suggested by Da Silva and De Souza Filho (2007) and Moir (2010) to analyse the performance of the seed potato chain in Ethiopia. The chain performance drivers are the enabling environment, technology, market structure, chain coordination, managing business operations, and inputs.

---

[7] To differentiate between local and improved potato varieties we used the definitions of Kaguongo *et al.* (2008). According to these authors, local potato varieties are potato varieties grown by farmers, whose origin is unknown, or varieties released by the national agricultural research system (NARS) but which have been out with the farmers for more than 35 years without being cleaned up for diseases. Improved potato varieties are those varieties that have been developed or cleaned up for diseases by CIP in collaboration with national research stations since 1970 and which are considered to be superior in qualities such as yield, resistance to diseases, dormancy period, maturity period or taste, compared with local varieties.

5.2.1 The enabling environment

The enabling environment comprises policies, institutions and support services that form the general setting under which enterprises are created and operate. Depending on the way it is arranged, the enabling environment can either support or harm the performance of an agri-commodity chain. The enabling environment components considered in this study are: provision of business development services such as training and extension to seed potato growers and the key providers of these business development services such as non-governmental organisations (NGOs), governmental organisations and private sector; research and development institutions supporting technology transfer and quality control; and laws and regulations regarding land tenure and access to land.

5.2.2 Technology

Technologies are essential determinants of value chain performance through their association with production, processing and distribution operations along the chains. Technologies include the methods, processes, facilities and equipment used in chain operations plus those applied in research and development. The components of technologies considered as chain performance drivers in this study are: availability and adoption of improved varieties, use of improved/diffused light store (DLS) storage methods by farmers, processing equipment, and transportation facilities.

5.2.3 Market structure

Evaluation of market structure shows the extent to which markets are competitive or whether they are characterised as oligopolistic or monopolistic markets. Market structure has impact on the performance of individual firms (business operations) at each stage of the value chain. The components of the market structure considered in this study include number and type of buyers.

5.2.4 Chain coordination

Chain coordination refers to the harmonisation of the physical, financial and information flows and of property right exchange along a chain (Moir, 2010; Da Silva and De Souza Filho, 2007). Well-functioning coordination facilitates planning and synchronising such flows and exchanges among the chain's different echelons, thus promoting organisational efficiencies (Moir, 2010; Da Silva and De Souza Filho, 2007). These, in turn, should translate into lower systemic costs, better consumer responsiveness and increased overall competitiveness. Coordination is affected by governments and/or organisations that can play a direct role in establishing or fostering public and private sector strategies and policies of interest to a particular chain. The evaluation of coordination should concentrate on the mechanisms that govern transactions among chain participants and on the effectiveness of such mechanisms in promoting the harmonisation of the physical, financial and information flows and of property right exchange along a chain. Components of coordination considered in this study are degree of chain organisation measured by the presence of farmers' primary cooperatives and chambers of commerce that play a direct role in establishing or fostering public and private sector strategies and policies of interest to the development of the seed potato value chain.

5.2.5 Managing business operations or farm management

The ability of individual firms to efficiently allocate resources, respond to consumer needs and adapt to market changes is to a great extent a function of its managerial power. Components of managing business operations considered in this study are efficient allocation of resources, response to consumer needs, and adaptation to market changes.

5.2.6 Inputs

The availability and cost of the main inputs (land, labour and capital) in the different segments of a chain directly affect performance at every stage. Inputs considered in this study are availability and cost of land, labour, seed, fertilisers, and pesticides.

5.3 Methodology

The performance analysis first conducted a mapping of the seed potato value chains currently existing in Ethiopia followed by a description of the chain using the chain performance drivers. In order to map the seed potato value chain and to describe and analyse its performance and improvement options, a literature review has been carried out, secondary data has been collected (for instance from local seed business and Ethiopian Central Statistical Agency reports), and key informants have been interviewed over the course of 2008-2013, including experts from national and international agricultural research and development institutions such as the Ethiopian Institute of Agricultural Research, the International Potato Center (CIP) and Wageningen University. This study considers only components of the chain performance drivers relevant to the Ethiopian seed potato value chain. These components also vary depending on the actor group.

5.4 Results: performance analysis

5.4.1 Mapping and description of seed potato chain actors

The seed potato value chains in Ethiopia were mapped in such a way that they encompassed actors in three seed potato systems. The three seed potato systems operating within Ethiopia are informal, alternative and formal (Hirpa *et al.*, 2010). The informal seed potato system is the seed potato system in which tubers to be used for planting are produced and distributed by farmers without any regulation. The alternative seed potato system is the seed potato system that supplies seed tubers produced by local farmers under financial and technical support from NGOs and breeding centres. In Ethiopia there are community-based seed supply systems which are undertaken by the community with technical and financial assistance of NGOs and breeding centres. In the formal seed potato system seed tubers are produced by licensed private sector specialists and cooperatives. An integrated map of the three seed potato systems is presented in Figure 5.1. In the integrated map, seed potato flows among the main actor groups taking part in the three seed potato systems. In the *informal* seed system, the main actor groups of the seed potato value chain are ware growers saving seed for sale and own use, ware growers who occasionally buy seed tubers to renew their seed stock, ware growers who buy seed tubers every season, and ware/seed traders. The ware growers are smallholder farmers producing potato for home consumption and cash.

In the *alternative* seed system, the main actors are breeding centres, NGOs, governmental organisations, and organised seed potato growers. The activities of the breeding centres are to develop potato varieties and to supply basic seed of these varieties to smallholder farmers and commercial seed growers. Holetta Agricultural Research Centre (HARC) and Haramaya University (HU) are the most important actors in developing potato varieties and supplying basic seed of improved varieties

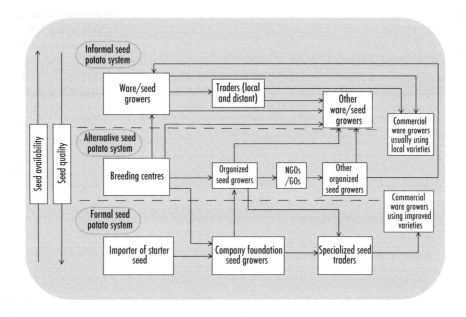

Figure 5.1. An integrated map of seed potato systems (delineated by dotted lines) in Ethiopia. NGO = non-governmental organisation; GO = governmental organisation.

in Ethiopia. Organised seed growers are smallholder farmers organised in farmers' research groups (FRGs) or farmers' field schools (FFSs), set up by research centres (especially Holetta Agricultural Research Centre). Currently, most of the FRGs and FFSs are transformed into seed producers' cooperatives. According to Hirpa (2013), there are more seed potato producers' cooperatives in the central and southern potato producing areas than in the eastern and north-western areas. The cooperatives in the central areas have more experience with producing seed than the cooperatives in the southern areas.

The actors in the *formal* seed system are modern commercial seed growers and ware/seed growers. Solagrow Private Limited Company (PLC) is the only modern commercial seed potato producer in the formal seed potato system. The PLC supplied seed potatoes to two seed potato grower cooperatives in the districts of Ambo and Doba. The cooperative in Ambo had 30 members (ISSD, 2011). These cooperatives out-grew seed potato for Solagrow PLC, which means that Solagrow PLC was the seed supplier and the seed buyer. The PLC plans to export seed potatoes to neighbouring countries. Ware/seed growers are domestic farmers or farmers in the importing countries.

Table 5.1. Main actors of the present seed potato value chain in the main potato growing areas of Ethiopia.[1]

Actor groups	Different potato producing areas[2]			
	Central	Eastern	Northern and north-western	Southern
Ware/seed growers	0.20 million	0.04 million	0.58 million	0.46 million
Breeding centres	HARC	HU	AARC, DARC	AwARC
Organised seed growers	16 SPGCs that comprised 744 members	One SPGC that comprised 40 members and 70 out-growers	One FRG that comprised 30 members and one SPGC that comprised 34 members	15 newly established SPGCs that comprised about 300 members
NGOs, projects and governmental organisations	DBARD	FAO, SHA, DBARD	CIP, World Vision, TDA, DBARD	CIP, GOAL Ethiopia, Vita, DBARD
Modern seed growers	Solagrow PLC	No	No	No

[1] AARC = Adet Agricultural Research Centre; AwARC = Awassa Agricultural Research Centre; CIP = International Potato Center; DARC = Debreberhan Agricultural Research Centre; DBARD = District Bureau of Agriculture and Rural Development; FRG = Farmers' Research Group; HARC = Holetta Agricultural Research Centre; HU = Haramaya University; SHA = Self-Help Africa; SPGC = Seed Potato Growers' Cooperative; TDA = Tigray Development Association.

[2] Central comprises east Shewa zone, north Shewa zone, west Shewa zone, south-west Shewa zone and Gurage zone; eastern comprises east Hararge zone, west Hararge zone, Harari regional state and Dire Dawa; northern and north-western comprises Tigray and Amhara regional states; southern comprises all potato growing zones in SNNPR except Gurage zone, Arsi zone and west Arsi zone.

The distribution of the actors across the major potato growing areas of Ethiopia is given in Table 5.1. The formal seed system is found only in the central potato growing area of Ethiopia.

5.4.2 Evaluation of actors against value chain performance drivers

In this section, we describe how each chain actor group is affected by the different chain performance drivers. As indicated above, the following chain performance drivers are used: enabling environment, technology, market structure, chain coordination, firm management and input.

5.4.2.1 Enabling environment

In the informal seed system, the District Bureau of Agriculture (DBARD) is the institution responsible for providing extension services for all crops including potato. Trained Development Agents (DAs) provide the extension service to farmers. According to Davis *et al.* (2010), there were 8,489 farmer training centres in Ethiopia in 2009, staffed with 45,812 DAs. The technical support that the farmers obtained from the agricultural DAs, however, is limited because (1) DAs were more involved in input supply, collecting tax and loan repayment than in providing technical support to farmers (Kassa, 2003) and (2) the DAs do not have sufficient technical knowledge (Kassa and Degnet, 2004; Davis *et al.*, 2010). In Ethiopia, less emphasis has been given to technical assistance on potato compared with technical assistance on major cereal crops like maize, teff, barley and sorghum. Potato is not among the priority crops for which an extension package is developed and implemented by the government of Ethiopia. CIP, the agricultural research centres and the NGOs have been undertaking the extension services on potato. The extension efforts by CIP and the agricultural research centres have focused on replacement of local potato varieties by improved potato varieties (Gebremedhin *et al.*, 2008). Little effort has been made to improve husbandry of local potato varieties despite their importance in total potato production. The local varieties comprise more than 90% of potato production (Gildemacher *et al.*, 2009; Hirpa *et al.*, 2010). In Ethiopia there is no institution responsible for the control of quality of seed potato. However, CIP in partnership with the Ministry of Agriculture and the agricultural research institutes has developed a quality declared seed system for potato and sweet potato. The quality declared system is at the stage of implementation by the Ministry of Agriculture but is threatened by the ubiquitous quarantine disease bacterial wilt (*Ralstonia solanacearum*).

The agricultural research centres, Haramaya University and Solagrow PLC are sources of seed potato for chains operating in the alternative and formal seed systems (Figure 5.1). Among the agricultural research centres, Holetta has been developing new potato varieties in collaboration with CIP. The role of Holetta is to test CIP potato clones for Ethiopian growing conditions. Other research centres such as Adet, Debre Berhan and Awassa demonstrate new potato varieties developed by Holetta Agricultural Research Centre. According to Haverkort *et al.* (2012), in 2012 all research centres together had 15 researchers working on potato and an annual budget of Ethiopian Birr (ETB) 196,000 (about USD 10,000). Among the NGOs in the alternative seed system, CIP with its project entitled 'Better Potato for a Better Life' has been working to improve seed supply systems in Tigray and Southern Nations, Nationalities and Peoples Regions.

In the alternative seed system, breeding centres in collaboration with DBARD provide training on seed potato production to the organised seed growers. Some of the breeding centres are also involved in seed quality control and marketing. For example, potato researchers in Holetta Agricultural Research Centre have been supervising seed potato fields of members of cooperatives to examine the level of late blight and bacterial wilt infestations. Some cooperatives in the district of Welmera have committee members who supervise seed potato fields of member farmers for disease infestation (especially for bacterial wilt). In the alternative seed system, the role of NGOs is to distribute seed potatoes from the organised seed growers to other organised or individual farmers and provide technical assistance to farmers (for instance, assist farmers in constructing diffused light store).

Solagrow PLC, Holetta Agricultural Research Centre and DBARD are the institutions that provide support to the chain in the formal seed system. The PLC provides inputs and technical advice to the seed potato out-growers to ensure they produce seed of the required quality. Holetta Agricultural Research Centre has been providing seed potatoes in the formal seed system to smallholder farmers and commercial seed growers. For instance, Holetta Agricultural Research Centre has provided seed potatoes to Solagrow PLC. The land administration section of DBARD is responsible for providing land for seed potato production to produce seed potatoes.

5.4.2.2 Technology

In the informal seed potato system, the potato growers predominantly used local potato varieties to produce ware and seed potatoes. Seeds of improved potato varieties were unavailable to the majority of these potato growers. In this seed system, ware/ seed potato growers used local storage methods, such as leaving in the soil (postponed harvesting), storing in a local granary, in jute sacks, or in a bed-like structure to store seed potatoes (Gildemacher *et al.*, 2009; Hirpa *et al.*, 2010). Local storage methods are largely sub-optimal because seed potatoes stored in local storage have fewer, longer and weaker sprouts that have low vigour at planting (i.e. are physiologically too old) (Gildemacher, 2012). In the alternative seed system, seed growers used the DLS method. According to Hirpa *et al.* (2012), use of DLS was perceived by seed growers in districts of Jeldu and Welmera to significantly improve seed yield compared with use of local storage methods.

With regard to transport, the means used in the informal and alternative seed systems are inappropriate. According to Hirpa *et al.* (2010), seed potatoes are usually transported by pack animals, tied by ropes on their backs, which could cause bruising.

In the formal seed system, seed potato is packed in boxes and transported by lorries. Seed damage during transportation is deemed to be very low. In the formal seed system, Solagrow PLC has two large seed storage facilities located at Hidi and Wonchi. It has its own plant and molecular laboratory (ELISA and qPCR) which is important to certify the phytosanitary quality of seed potatoes (Solagrow PLC, 2011). The company also has a seed tuber grading facility and modern ICT facilities to communicate with its customers.

5.4.2.3 Market structure

In the informal seed system, there are many ware/seed sellers and buyers. Price is the main determinant of transaction. In the alternative seed system, the major buyers of the seed produced by the organised seed growers are NGOs. Government organisations also buy small amounts of seed from organised seed growers. As a result of these inspirational buyers the seed potato market is highly distorted. Demand for seed of improved potato varieties from smallholder ware growers is very low because of the high price of the seed. Moreover, some ware potato growers are not aware of the benefits of using seed tubers of improved potato varieties. Abebe *et al.* (2013a) found that farmers in the Shashamene region preferred local potato varieties, such as Netch Abeba, above improved potato varieties such as Jalene and Gudene because of better agronomic characteristics and cooking quality of the former. The formal seed system is in an incipient stage. In this seed potato system, there is only one seed potato producer (i.e. Solagrow PLC). The PLC contracts smallholder farmers to produce seed potato and buy back the seed. The company grades and certifies seed, and sells it to ware growers.

5.4.2.4 Chain coordination

There is no formal institution that attempts to coordinate the informal seed system. In this seed potato system, ware/seed growers supply ware/seed to local markets and ware/seed buyers buy what is available in the market. Farmers obtain information on potato production and marketing from a variety of sources. According to Gildemacher *et al.* (2009), most potato growers in Ethiopia used their own experience to select seed potato (57.3%), to enrich their soils (63.7%), to undertake general crop husbandry (59.2%), post-harvest handling (62.9%), and marketing (70.4%). These authors also reported that farmers' own communities were the major sources of information about potato varieties (58.7%) and crop protection (33.7%).

In the alternative seed system, breeding centres are coordinated at national level. According to Haverkort *et al.* (2012), the existence of national coordination helped the breeding centres to use their research resources efficiently. Some NGOs formed a consortium which helps them to coordinate potato development activities in Ethiopia. These NGOs are Catholic Relief Services (CRS), FAO Ethiopia, Food for the Hungry (FH), GOAL Ethiopia, World Vision Ethiopia, ChildFund, Concern Worldwide and International Medical Corps (IMC). The consortium is organised and established by

FAO and aimed at implementing a project titled 'Disaster Risk Management – Root and Tuber Crops Response Intervention in SNNP Region' (G. Solomon, CIP-Addis Ababa, personal communication). The aim of the consortium is to achieve a wide geographic coverage and to coordinate and harmonise activities.

In the alternative seed system the market for seed potatoes produced by organised seed growers is a major problem because of unreliable demand. Buyers only buy seed when they have money. According to key informants, a decision on the area of land to be allotted for seed potato production by farmers in a certain season depends on the price of seed in the preceding season. A season with high prices usually leads to oversupply of seed potato in the subsequent season resulting in a very low price.

In the formal seed system, there is coordination among actors in the seed value chain. For instance, there is a contract between the growers and Solagrow PLC, which supplies the seeds to and buys the tubers from the growers (ISSD, 2011). Based on the contractual agreement, the PLC pays different prices for different grades of seed potatoes.

5.4.2.5 Seed potato farm management

In the informal seed system, ware/seed growers use suboptimal farm management practices (Hirpa *et al.*, 2012). Seed potato management practices that enhance the quality of potatoes are not practised by most of the growers. These practices are the following: use of appropriate seed tuber size; healthy seed; positive selection; rotation and allocation of a separate plot for seed production. According to Hirpa *et al.* (2010), in 2007 only 13% of the farmers in the district Degem, 15% of the farmers in the district Jeldu in the central area and 8% of the farmers in the district Banja in the north-western area produced seed potatoes by positive selection, whereas 1% of the farmers in district Degem, 14% of the farmers in the district Jeldu and 6% of the farmers in the district Banja produced seed potato on separate plots. In the southern area there is no practice of positive selection or use of separate plots for the production of seed tubers. According to Mulatu *et al.* (2005), farmers in the eastern part of Ethiopia usually do not produce seed tubers on separate plots. In this part of the country there is no positive selection either.

Solagrow PLC is the only modern and licensed seed potato growing PLC in Ethiopia. The PLC follows modern farm management practices to multiply seed potatoes. Its technical experts assist the out-growers to follow modern farm management practices.

5.4.2.6 Inputs

In the informal seed system, the majority of the farmers use seed tubers of local varieties to produce ware/seed. Ware/seed growers saving seed are the main suppliers of seed potatoes of local potato varieties. Self-supply and local markets are the major sources of seed potato of local varieties. Small amounts of seed potato of improved varieties are supplied to this seed system by breeding centres and organised seed growers. According to Gildemacher *et al.* (2009), of the total seed potato used in Ethiopia, seed tubers of improved potato varieties comprised only 1.3%. Agrochemicals are supplied by the DBARD[8] and traders in the vicinity, in order of importance. According to CSA (2011), of the total land allotted for potato cultivation in 2010, fertilisers (diammonium phosphate (DAP) and urea), fungicide and seed from improved potato varieties were used on 50%, 25% and 0.5% of the land, respectively. In 2010, on average 80 kg/ha of fertiliser was used for potato production.

In the alternative seed system, breeding centres supply basic seed of the improved potato varieties they developed to cooperatives of seed growers who produce seed potatoes from this basic seed. The members of these cooperatives produce seed potato on their individual plots. Governmental organisations and NGOs bought seed tubers from cooperatives and distributed them to farmers in distant areas. In this seed system, DBARD is the main supplier of agrochemicals to the organised seed growers. Private traders also supply agrochemicals. In the alternative seed system a high amount of fertiliser is used for seed potato production. According to Hirpa *et al.* (2012), in 2010 seed growers in Jeldu and Welmera used 245 kg/ha and 238 kg/ha of fertilisers (DAP and urea together) respectively, to produce seed tubers of improved potato varieties.

The only potato company in the formal system, Solagrow PLC, obtains basic seed from the Dutch seed potato company HZPC Holland BV. The company also obtains seed from Holetta Agricultural Research Centre. According to Haverkort *et al.* (2012), the seed potato obtained from Holetta Agricultural Research Centre comprises 20% of the total seed used by the PLC. The PLC sells seed, and other inputs to the organised seed growers (Abebe *et al.*, 2013b). These organised growers have access to irrigation water through irrigation systems constructed by the government.

5.5 Improvement options for the seed potato value chains

5.5.1 Enabling environment

Currently, the DAs that should provide extension services to farmers may have not enough technical knowledge on potato cultivation (cf. Davies *et al.*, 2010). They are also engaged in other activities such as input distribution, and collecting tax and loans. Provision of on-the-job-training could improve the technical knowledge of the DAs. Besides, they should be fully engaged in the provision of technical assistance to farmers.

Potato is an important food security crop, but is not among the crops being prioritised for expansion through formal agricultural extension programmes. Therefore, the

---

[8] Fertilisers distributed to the farmers through the DBARD are imported by input supplying cooperatives.

government has to give more emphasis on rendering extension services to stimulate potato development.

There is also insufficient budget for potato research. As a result the number of staff engaged in research is low. For instance, only 15 researchers were engaged in potato research in Ethiopia in 2012 (Haverkort *et al.*, 2012), a number well below 1.5% of the total number of agricultural research staff in 2008. In 2008, the total number of agricultural staff was 1,318 (Flaherty *et al.*, 2010). Therefore, a larger budget has to be allotted to increase the number of staff in potato research.

In the informal and alternative seed systems, seed potato quality is a problem, whereas there is no quality control on the seed produced by the majority of seed growers during production and storage. However, seed producer cooperatives in South and Central Ethiopia have developed a quality control system. For instance, two seed growers' cooperatives in Welmera district have committees that supervise members' seed potato fields for diseases, such as bacterial wilt and late blight. These committee members obtained training from staff of Holetta Agricultural Research Centre. Building technical capacity of the quality control committee through training and encouraging other cooperatives to adopt this system will contribute to the enhancement of potato production. According to Hirpa *et al.* (2010), design of a quality control system is important to improve the alternative seed potato system in Ethiopia.

In Ethiopia, there is no formal institution responsible for seed potato quality control and certification. However, CIP and the Ministry of Agriculture (MoA) have developed a seed potato quality control system which is in the process of being institutionalised.

5.5.2 Technology

In the informal seed system, all ware/seed potato growers use local storage methods such as leaving in the soil (postponed harvesting), local granary, jute sacks, and bed-like structure to store seed potatoes. These methods are largely sub-optimal because seed potatoes stored in local storage have fewer, longer and weaker sprouts that have low vigour at planting and subsequently produce a low yield. The use of DLS, a good seed potato storage method, is limited to seed growers that produce seed potatoes of improved varieties. DLS can, however, also be used to store seed tubers of local varieties. To this end ware/seed growers of local varieties have to be informed about the importance of DLS. Seed quality loss also occurs during harvesting and transportation. Farmers have to be trained on the method that helps them to minimise seed tuber quality loss during harvesting and transportation. Farmers have to use farm or transport equipment that does not bruise, cut or pierce tubers during harvesting or transport.

5.5.3 Market structure

In the alternative seed system, the major buyers of the seed potatoes are NGOs. There is no involvement of the private sector in trade in seed potatoes of improved varieties. Involvement of private sector in seed potato trading is highly important. Therefore, linkage between traders and seed producers has to be created to change the existing monopsonistic seed potato market.

5.5.4 Chain coordination

In the informal seed potato system, there is no coordination within actor groups and among different chain actors in the ware/seed value chain. In the alternative seed system, there is some coordination among some actors. For example, the breeding centres are coordinated at national level; some NGOs operating in the southern part of Ethiopia have also formed a consortium that coordinates and harmonises their activities. There are also seed growers' cooperatives in central potato growing areas of Ethiopia that search markets for seed potatoes produced by cooperative members. However, these all operate at the level of the individual actors and little coordination exists across the chain actors within a particular seed system. Creating commodity associations and other forms of trader groups that can play a direct role in establishing or fostering public and private sector strategies and policies of interest to the seed potato chain could improve the supply of and demand for quality seed potatoes. Contracts can be one form of a chain coordination mechanism. The seed value chain in the formal system is coordinated through contracts. There is a good coordination between the chain actors in the formal system, as Solagrow PLC has a market, input and advisory service contract with its out-growers (ISSD, 2011).

5.5.5 Seed potato farm management

Good farm management can enhance yield and quality of seed potatoes produced by ware/seed growers. For example, a positive seed potato selection study undertaken in Kenya showed that positive seed potato selection increases the yield of potato by ware growers by about 34% (Gildemacher *et al.*, 2012). The current practice in the informal seed system is suboptimal and needs improvement. Advising farmers to follow good management practice can help to enhance the seed potato management practices by farmers and will result in better seed tuber health and better seed vigour.

5.5.6 Inputs

In the informal seed system, seed potato is traded as part of general ware/seed and other agricultural products. There is little differentiation between ware and seed and between different potato varieties. In this seed system, there is no label that indicates the variety name or origin of the potato. Tuber size is one criterion for potato tubers to be used for seed or ware: usually small-sized tubers are used for seed and large-sized tubers are used for ware (Endale *et al.*, 2008; Gildemacher *et al.*, 2007; Mulatu *et al.*,

2005). Small-sized tubers may have two problems. The first one is delayed emergence and low sprout vigour and number because of low food reserve (Lommen, 1994; Lommen and Struik, 1994) and consequently low yield and tuber number (Lommen and Struik 1994, 1995). The second is that they might be a progeny of an infected mother plant and thus infected by diseases, because infected mother plants usually give small tubers (Struik and Wiersema, 1999). In Ethiopia, the use of small potato tubers as seed might have contributed to the building up of a high level of disease infection especially in the local varieties. Therefore, potato growers have to be advised not to use small-sized tubers for seed. In this seed potato system, the use of agrochemicals is also very low; that needs to be improved as well.

In Ethiopia, the use of seed tubers of improved potato varieties by potato growers is very low. One cause is unavailability of seed tubers of improved potato varieties to the majority of the growers (Hirpa *et al.*, 2010; International Potato Center, 2011). Availability of seed tubers of improved varieties has to be increased. This can be achieved by encouraging the private sectors to participate in seed production and supply. A good supply of seed tubers of improved potato varieties also requires involvement of smallholder farmers in the seed potato production. According to Hirpa *et al.* (2012), high costs related to adoption of improved potato varieties plus their advised production methods could be one of the causes for the low uptake of improved varieties. This problem can be solved by availing to smallholder farmers, production methods that are affordable and at the same time can enable them to produce seed tubers with reasonable yield and quality.

5.6 Conclusions

This chapter analyses the performance of the seed potato value chains with respect to their ability to supply quality seed tubers to seed potato systems in Ethiopia using the following value chain performance drivers: enabling environment, technology, market structure, chain coordination, farm management, and inputs. All seed potato value chains experience problems with respect to supplying quality seed tubers and thus need improvements. Poor quality seed tubers supplied by the value chains in the informal seed system are related to a poor enabling environment such as a low quality technical support and lack of seed quality control system. Other causes of poor quality seed tuber in this seed system are the use of sub-optimal storage and transportation technologies, sub-optimal farm management practices, use of poor quality seed tubers and use of low level of inputs (fertilisers and crop protection chemicals). The results show that in order to improve the quality of tubers in the informal seed system, it is important to: (1) improve the technical knowledge of DAs; (2) use improved storage methods; (3) advise farmers and transporters how to minimise quality loss during harvesting and transportation; (4) advise farmers to follow improved farm management practices; (5) inform farmers to use seed tubers of the right physiological age; (6) advise farmers not to use small-sized tubers for seed; and (7) advise and facilitate farmers to improve the amount of fertilisers and crop protection chemicals.

The major problem of the chains in the alternative and formal seed potato systems is that they supply only a very small amount of seed tubers from improved varieties, mainly because of the negligible role of the private sector in producing and supplying seed tubers of improved varieties. Moreover, the seed tubers supplied by the chains in the alternative seed system are not of a standard quality mainly because of absence of quality control system. A major portion of the seed potato produced through alternative seed potato system is bought by institutions which created a distortion in the seed potato market. There is a need to establish or strengthen institutions that have a mandate to support and encourage private sector involvement and the establishment of quality control systems in the production and supply of seed tubers in the alternative and formal seed systems.

References

Abebe, G.K., Bijman, J., Pascucci, S., and Omta, O., 2013a. Adoption of improved potato varieties in Ethiopia: the role of agricultural knowledge and innovation system and smallholder farmers' quality assessment. Agricultural Systems 122: 22-32.

Abebe, G.K., Bijman, J., Kemp, R., Omta, O. and Admasu T., 2013b. Contract farming configuration: smallholders' preferences for contract design attributes. Food Policy 40: 14-14.

CSA (Central Statistical Agency of Ethiopia), 2011. Agricultural sample survey: report on area and production of crops. CSA, Addis Ababa, Ethiopia.

Da Silva, C.A. and De Souza Filho, H.M., 2007. Guidelines for rapid appraisals of agrifood chain performance in developing countries. FAO, Rome, Italy. Available at: http://www.fao.org/3/a-a1475e.pdf.

Davis, K., Swanson, B., Amudavi, D., Mekonnen, D.A., Flohrs, A., Riese, J., Lamb, C. and Zerfu, E., 2010. In-depth assessment of the public agricultural extension system of Ethiopia and recommendations for improvement. IFPRI discussion paper 01041. Available at: http://tinyurl.com/qzw2rb5.

Endale, G., Gebremedhin, W. and Lemaga, B., 2008. Potato seed management. In: Gebremedhin, W., Endale, G., Lemaga, B. (eds.) Root and tuber crops: the untapped resources. Ethiopian Institute of Agricultural Research, Addis Ababa, Ethiopia, pp. 53-78.

FAO, 2008. Potato world: Africa – international year of the potato 2008. Available at: http://www.fao.org/potato-2008/en/potato/.

Flaherty, K., Kelemework, F. and Kelemu, K., 2010. Recent developments in agricultural research. IFPRI and EIAR. Available at: http://www.asti.cgiar.org/pdf/Ethiopia-Note.pdf.

Gebremedhin, W., Endale G. and Lemaga, B., 2008. Potato variety development. In: Gebremedhin, W., Endale G. and Lemaga B. (eds.) Root and tuber crops: the untapped resources. Ethiopian Institute of Agricultural Research, Addis Ababa, Ethiopia, pp. 15-32.

Gildemacher, P., Demo, P., Kinyae, P., Wakahiu, M., Nyongesa, M. and Zschocke, T., 2007. Select the best; positive selection to improve farm saved seed potatoes. Training manual. CIP, Lima, Peru.

Gildemacher, P., Demo, P., Barker, I., Kaguongo, W., Gebremedhin, W., Wagoire, W.W., Wakahiu, M., Leeuwis, C. and Struik, P.C., 2009. A description of seed potato systems in Kenya, Uganda and Ethiopia. American Journal of Potato Research 86: 373-382.

Gildemacher, P., 2012. Innovation in seed potato systems in Eastern Africa. PhD thesis, Wageningen University. KIT Publishers, Amsterdam, the Netherlands.

Gildemacher, P.R., Leeuwis, C., Demo, P., Borus, D., Schulte-Geldermann, E., Kinyae, P., Mundia, P., Nyongesa, M. and Struik, P.C., 2012. Positive selection in seed potato production in Kenya as a case of successful research-led innovation. International Journal of Technology Management and Sustainable Development 11: 67-92.

Haverkort, A., Van Koesveld, F., Schepers, H., Wijnands, J. Wustman, R. and Zhang, X., 2012. Potato prospects for Ethiopia: on the road to value addition. Wageningen UR, Wageningen, the Netherlands. Available at: http://edepot.wur.nl/244969.

Hirpa, A., 2013. Economic and agronomic analysis of seed potato supply chain in Ethiopia: Thesis, Wageningen University, Wageningen, the Netherlands.

Hirpa, A., Meuwissen, M.P.M., Tesfaye, A., Lommen, W.J.M., Lansink, A.O., Tsegaye, A. and Struik, P.C., 2010. Analysis of seed potato systems in Ethiopia. American Journal of Potato Research 87: 537-552.

Hirpa, A., Meuwissen, M.P.M., Van der Lans, I., Lommen, W.J.M., Oude Lansink, A.G.J.M., Tsegaye, A. and Struik, P.C., 2012. Farmers' opinion on seed potato management attributes in Ethiopia: a conjoint analysis. Agronomy Journal 104: 1413-1423.

International Fund for Agricultural Development (IFAD), 2010. Rural poverty report 2011: new realities, new challenges: new opportunities for tomorrow's generation. IFAD, Rome, Italy.

International Potato Center (CIP), 2011. Roadmap for investment in the seed potato value chain in eastern Africa. CIP, Lima, Peru, 27 pp.

Integrated Seed Sector Development (ISSD), 2011. Local seed business newsletter 8. Available at: http://tinyurl.com/nbngcoc.

Kaguongo, W., Gildemacher, P., Demo, P., Wagoire, W., Kinyae, P., Andrade, J., Forbes, G., Fuglie, K. and Thiele, G., 2008. Farmer practices and adoption of improved potato varieties in Kenya and Uganda. Social Sciences Working Paper 2008-5. International Potato Center (CIP), Lima, Peru, 85 pp.

Kassa, B., 2003. Agricultural extension in Ethiopia: the case of participatory demonstration and training extension system. Journal of Social Development in Africa 8: 49-83.

Kassa, B. and Degnet, A., 2004. Challenges facing agricultural extension agents: a case study from South-Western Ethiopia. African Development Bank, Abidjan, Ivory Coast.

Lommen, W.J.M., 1994. Effect of weight of potato minitubers on sprout growth, emergence and plant characteristics at emergence. Potato Research 27: 315-322.

Lommen, W.J.M. and Struik, P.C., 1994. Field performance of potato minitubers with different fresh weights and conventional tubers: crop establishment and yield formation. Potato Research 37: 301-313.

Lommen, W.J.M. and Struik, P.C., 1995. Field performance of potato minitubers with different fresh weights and conventional tubers: multiplication factors and progeny yield variation. Potato Research 38: 159-169.

Moir, B., 2010. Value chain theory and application. In: Strengthening potato value chains: technical and policy options for developing countries. FAO, Rome, and Common Fund for Commodities, Amsterdam, the Netherlands.

Mulatu, E., Osman E.I. and Etenesh, B., 2005. Improving potato seed tuber quality and producers' livelihoods in Hararghe, Eastern Ethiopia. Journal of New Seeds 7: 31-56.

Solagrow, P.L.C., 2011. About solagrow. Available at: http://www.solagrowplc.com/about-solagrow.

Spielman, D.J., Byerlee, D., Alemu, D. and Kelemwork, D., 2010. Policies to promote cereal intensification in Ethiopia: the search for appropriate public and private roles. Food Policy 35: 185-194.

Spielman, D.J., Kelemwork, D. and Alemu, D., 2011. Seed, fertilizer, and agricultural extension in Ethiopia. In: Dorosh, P. and Rashid S. (eds.) Food and agriculture in Ethiopia: progress and policy challenges. University of Pennsylvania Press, Philadelphia, PA, USA, pp. 84-122.

Struik, P.C. and Wiersema, S.G., 1999. Seed potato technology. Wageningen Academic Publishers, Wageningen, the Netherlands.

USAID, 2011. Feed the future: Ethiopia multi-year strategy (2011-2015). Available at: http://tinyurl.com/h57taju.

6. Diverging quality preferences along the supply chain: implications for variety choice by potato growers in Ethiopia

G.K. Abebe[1], J. Bijman[2], S. Pascucci[2], S.W.F. Omta[2] and A. Tsegaye[3]*

[1]*American University of Beirut, Department of Agricultural Sciences, P.O. Box 11-0236, Riad El Solh, Beirut 1107 2020, Lebanon;* [2]*Wageningen University, Management Studies, P.O. Box 8130, 6700 EW Wageningen, the Netherlands;* [3]*Addis Ababa University, P.O. Box 1176, Addis Ababa, Ethiopia; gumataw@gmail.com*

Abstract

Improving the introduction of new potato varieties requires aligning the preferences of all supply chain actors. In Ethiopia, the majority of ware potato growers source their seed from the informal supply system. Using a case study on specialised seed growers and a survey among ware potato growers and downstream chain actors, we explore the quality attributes that could influence the variety choices of farmers and downstream actors. Especially, we analysed the link between the seed and ware potato supply chains, farmers' evaluation of local and improved potato varieties, and quality differences between the local and improved varieties. We found that farmers' variety choices are well-aligned with traders' preferences but varieties supplied by the specialised seed potato growers are not well accepted by ware potato growers. As a result, ware potato farmers continue to grow local varieties, which are inferior in terms of production-related quality attributes, but superior regarding market-related quality attributes. The results imply that enhancing production-related quality attributes is not enough to induce farmers to accept new potato varieties. We recommend breeding institutes and seed potato growers to put more emphasis on market-related quality attributes to enhance choice alignment in the full potato chain.

Keywords: quality preferences, production related quality attributes, market related quality attributes, variety choice

6.1 Introduction

Vertical coordination in food chains has become increasingly important due to consumers' demand for a higher quality product, and because competition in national and global markets has shifted from price-based to quality-based (Henson and Jaffee, 2008; Swinnen and Maertens, 2007). In international food chains, retailers and large processors have taken up the coordination role of aligning decisions on quality, for instance, by imposing their own food quality and safety protocols (Narrod *et al.*, 2009). In domestic markets in developing countries, aligning quality preferences is more difficult as chain actors tend to operate independently.

Potato is increasingly demanded in developing countries because of population growth (Pretty *et al.*, 2003) and growth in fast food restaurants (Stewart *et al.*, 2004), but also because consumers perceive potato as a healthy food (Jemison *et al.*, 2008). Ethiopia is experiencing both population growth and a rise in the number of fast food restaurants (Gildemacher *et al.*, 2009a). Aligning quality decisions is usually more difficult in the potato chain than in other chains. First, production of potato is more complex than of other crops, because of the difference in agro-ecological requirements for seed and ware potato production[1]. Second, producers of seed potatoes are often disconnected from producers of ware potatoes because ware potato growers use either their own (farmer-saved) seeds or seeds sourced from other ware potato growers, not from the specialised seed potato growers[2]. Third, lack of common quality grades and standards or formal mechanisms in trading quality declared seeds makes the alignment of quality decisions between seed potato growers and ware potato growers a challenge.

The objective of this paper is to provide insights in quality attributes influencing variety choices and alignment of quality preferences between different actors in the Ethiopian potato supply chain. We explore how differences in quality attribute preferences affect the choice for specific varieties. To do so, we address four separate but related questions: (1) How are the specialised seed potato growers connected to the ware potato growers? (2) How do ware potato farmers characterise the currently used potato varieties? (3) How do quality attributes of the improved seed potato varieties compare with the local varieties? (4) Which quality attributes influence ware potato growers' choice for specific varieties and how does this relate to the preferences of downstream actors?

While seed potato supply systems of developing countries have received much academic attention in recent years (Gildemacher *et al.*, 2009a,b; Hirpa *et al.*, 2010, 2012; Ortiz *et al.*, 2013; Schulte-Geldermann *et al.*, 2012), and have been the focus of much work by the International Potato Centre (CIP), much less attention has been paid to the linkage between the improved seed supply chain and the ware potato supply chain. This paper aims to combine economics/marketing and agronomic concepts to answer the above research questions.

The paper is organised as follows. The next section presents an overview of the Ethiopian potato sector and a conceptual framework, followed by a materials and methods section. The final sections provide the results and discussion and conclusion.

---

[1] Seed potatoes are often grown in cooler areas to reduce the incidence of pests and diseases.

[2] Seed growers in this study refer to seed potato farmers who grow improved varieties released by the Ethiopian research institutes or by the International Potato Centre (CIP).

6.2 Overview of the Ethiopian potato sector

Potato is considered one of the spearheads of agricultural policy by the Ethiopian government because of its potential for food security and income generation. However, despite favourable government policies and good weather conditions, the productivity and market performance of the potato crop is still low. Over the last two decades, the increase in potato production has mainly come from area expansion (Figure 6.1). Modern varieties that could have enhanced smallholder productivity and commercialisation have not been widely adopted by the ware potato growers.

The Ethiopian Institute of Agricultural Research (EIAR) and CIP have put in substantial effort to improve the performance of the potato sector. With the help of international research organisations, EIAR has released 18 new potato varieties in the last two decades. Nonetheless, the adoption rate and area cultivated (by ware potato growers) with these varieties have been very low. Instead, smallholders continue to grow the local varieties.[3] The usual explanation for the low performance of the Ethiopian potato sector is poor seed supply systems (Gebremedhin et al., 2008). Three types of seed supply systems – informal, alternative and formal – co-exist in Ethiopia (Hirpa et al., 2010). Yet, more than 98% of potato growers source their seed from the informal seed market.

Why is the contribution of the specialised seed supply chain so low? Three explanations have been put forward. One explanation might be that there is not enough supply of improved potato seeds. A second explanation might relate to the demand-side: the demand for improved seeds by ware potato growers might still be low. A third explanation may be the lack of information about supply and demand conditions.

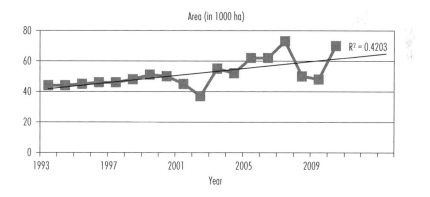

Figure 6.1. Area expansion and production of potatoes in Ethiopia between 1993 and 2010 (FAOSTAT, 2013).

---

[3] There are no indigenous potato varieties in Ethiopia. However, we use the term local to refer to those varieties for which no official record exists on when and by who they were first introduced.

In assessing these possible explanations, we used a case study approach to explore the supply-side conditions. In this case study, we focused on the specialised seed potato growers who are organised and supported by breeding centres and non-governmental organisations (NGOs) to multiply seed potatoes at commercial scale. The conditions on the demand side have been explored on the basis of a survey among ware potato growers and other actors on the downstream side of the potato chain.

6.3 Conceptual framework: quality preferences alignment

Several models can be found in the literature showing relationships between a good and quality. Lancaster (1966) was the first to define quality as a function of the characteristics of the good rather than the function of the good. That is, utility is provided by the individual attributes rather than the good *per se*. The Lancaster model of quality is based on objectively measurable properties of goods that are relevant to consumer choices. Preferences for a good can thus be determined by the relative weights assigned to the various attributes.

In a supply chain context, quality preferences of different chain actors may not be fully aligned due to several factors. First, different chain actors may perceive quality differently based on what they demand from and like about a particular product (Ruben *et al.*, 2007). Diverging perceptions (e.g. regarding the importance of specific quality attributes) among different actors can make the alignment of quality a complex and challenging task. Second, alignment of quality preferences requires some kind of coordination. Coordination is defined by Malone and Crowston (1994: 90) as 'the process of managing dependencies between activities'. Coordination combines information exchange and specific decision-making processes (Bijman and Wollni, 2009). Without information exchange, actors cannot know each other's preferences, and without a proper decision-making process, the preferences of different actors in a chain cannot be aligned. That is, the type of decision making process and the extent of information exchange can affect the alignment of quality preferences in a supply chain. Based on the above, we operationalise quality preferences alignment as the matching of preferences for particular quality attributes among different actors in a supply chain.

When a specific variety does not have the production- and market-related quality attributes as preferred by both farmers and traders, misalignment of preferences is present. The question then is which quality attributes influence growers' variety choice and how does this relate to the preferences of the downstream actors. In order to answer this question, it is necessary to characterise the production- and market-related quality attributes of the currently used varieties, and understand the quality differences across the varieties.

6.4 Materials and methods

Because different supply chain actors may define quality differently, we systematically analyse quality using two categories of quality attributes: production-related and market-related. Attributes like yield, disease tolerance, maturity period, drought resistance, and crop management intensity are production-related, because they determine production practices. Such attributes are expected to be used as criteria for selecting varieties for the production of seed potatoes. However, in the ware potato market, quality attributes such as tuber size, stew quality, cooking quality[4], colour, shape and shelf life are important. We call these attributes market-related because they determine sales options, and are more likely to define the criteria for selecting varieties for the production of ware potatoes.

For potato, variety type largely determines the intrinsic quality attributes (Howard, 1974; Jemison *et al.*, 2008). However, objective information on quality attributes of local varieties is lacking. In situations where no objective quality information is available, farmers' knowledge can be used to describe the quality attributes of different varieties (Cavatassi *et al.*, 2011). Thus, our study first addresses the question 'How do ware potato farmers characterise currently used local and improved potato varieties in terms of production- and market-related quality attributes?'

The main goal of characterising existing potato varieties was not to say whether a particular variety is good or bad. Rather, variety is generally considered as the main signal for potato quality, particularly between growers and their immediate buyers. But in Ethiopia variety names are short-lived and can simply die out; and hence, what seem more stable are the quality characteristics each variety possesses, not variety *per se*. Our research strategy was, therefore, designed in a way to effectively capture the most important quality attributes each variety possesses. Breeding centres can thus use this information to develop appropriate varieties that can satisfy the requirements of ware potato growers.

The extent of quality preferences alignment in the potato supply chain is measured by how the improved varieties grown by the specialised seed potato growers are linked to the demands of ware potato growers, and the extent to which these varieties are liked by downstream actors in the chain. To do so, the quality preferences of ware potato growers are measured by their decision to grow specific varieties over others, and the preferences of traders are measured by their preferences for specific quality attributes.

6.4.1 Case study

The focus of our case study was on seed growers who were organised in cooperatives and who were growing potato seed of improved varieties for commercial purpose.

---

[4] Stew quality refers to the taste of ware potatoes when they are boiled in a mix (i.e. a stew) with other vegetables. Cooking quality refers to the taste of the potato when consumed boiled without mixing with vegetables.

Such cooperatives were mainly located in the central highlands due to the good agro-ecological conditions (cool and high altitude). Cooperatives were initiated and supported by the Holeta Agricultural Research Centre (HARC). Based on the recommendation of HARC, we purposely selected two districts, Welmera and Jeldu. Subsequently, we selected 2 cooperatives from Welmera district and 1 cooperative from Jeldu district.

To gather data, personally administered semi-structured interviews were held with different stakeholders: leaders of the three cooperatives, eight cooperative members, and three experts from HARC and the Welmera district agricultural office. In addition, we observed the member farms and the storage facilities of the cooperatives. Parts of the questions to the cooperative leaders were related to quality improvement activities by seed growers, market access and support from external parties. Questions for members were related to production and marketing seed potatoes. The questions for the research centre and the agricultural development office were related to the type of support offered by the specialised seed growers. Ten storage facilities were visited, in two rounds, to observe how seed potatoes were stored and to get information on the time between entering of storage and sales of potatoes. The first visit was in November 2009, the time when potatoes were being harvested, and the second visit was in April 2010, the time when seed potatoes would be sold to ware potato growers.

6.4.2 Survey of farmers

We carried out the farmers' survey in the spring of 2011 among 350 ware potato farmers in the Rift Valley region of Ethiopia. Although potatoes can be produced in different parts of the country, our study focuses on the Rift Valley region for two main reasons. First, ware potato farmers in the region are the main suppliers of ware potatoes to the major cities of Ethiopia. For instance, the Shashemene spot market, in the centre of our study region, is considered as the main trade hub of ware potatoes (Emana and Nigussie, 2011; Tefera *et al.*, 2011). Second, to characterise the different potato varieties, it is necessary that the farmers in the survey have the same understanding of the currently used varieties. In Ethiopia, variety names lack standardisation and are often attached to local languages (Cavatassi *et al.*, 2011). Thus, focusing on one region avoids problems arising from confusion over variety names.

The ware potato farmers were randomly selected from the land ownership register obtained from the Office of Agriculture and Rural Development at the Shashemene, Shala, and Shiraro districts. The main objective of this part of the study was to characterise the different potato varieties, both the improved and local varieties.

Characterisation of the different potato varieties was carried out as follows. First, we identified seven potato varieties, and the classification was done by farmers, triangulated with information from agricultural agents. The varieties were: Aga Zer, Nech Abeba, Key Dinch, Key Abeba, Gudane, Jalene, and Bule. The latter three are improved varieties that were released by the Ethiopian Institute of Agricultural

Research (EIAR) while the others were local varieties for which no documentation was found on how and when these varieties were first introduced in the region. Second, to better understand the main characteristics each variety possesses, we classified quality into two dimensions: production-related and market-related.[5]

To understand the factors influencing their decision to grow a particular variety, ware potato farmers were asked the following question in relation to each variety: 'What was the main reason that led you to grow this variety in the previous season?' Farmers were asked to pick only one from nine different production- and market-related attributes described earlier. This would force farmers to reveal the most important reason for selecting one variety over others.

6.4.3 Survey of downstream actors

The main objective of this part of the survey was to understand the quality preferences of different actors in the downstream part of the potato supply chain. We carried out this survey in the summer of 2011 among 10 stationed wholesalers, 13 retailers, and 11 big hotels located in Addis Ababa. Because collecting wholesalers supply potatoes to the central market of Addis Ababa, we purposely selected the traders at the central market.[6] The question for the downstream actors was as follows: 'Please distribute 100 points over the different quality attributes that you may take into account when you are buying ware potatoes. Give the highest value to the most important quality attribute, the second highest value to the second most important quality attribute, etc.' Accordingly, the mean rank of each quality attribute would be computed to identify the highly ranked attributes, and compare it with the attribute considered most important by farmers in selecting a particular variety.

6.5 Results and discussion

6.5.1 Case study

The three seed potato cooperatives in Jeldu and Welmera areas were largely organised around farmers who used to participate in farmer research groups. HARC played a key role in the formation of the seed potato growers' cooperatives as part of its strategy to institutionalising the transfer of agricultural technologies to farmers. We observed two widely grown improved seed potato varieties, Jalene and Gudane, in the study area, which were released by HARC in 2002 and 2006 respectively. HARC was also the main provider of technical assistance to the specialised seed growers.

---

[5] Unobservable characteristics of potatoes were not considered in our analysis as information on these attributes is rarely conveyed in the supply chain.

[6] Collecting wholesalers are traders who are located close to ware potato farmers and they supply potatoes to stationed wholesalers. Stationed wholesalers are traders who buy potatoes from collecting wholesalers and distribute it to retailers, hotels and restaurants. Stationed wholesalers are located close to retailers and final consumers (thus at the central market in Addis Ababa).

This indeed had helped the growers to improve their seed quality and productivity. In addition to providing technical assistance, HARC served as an intermediary in the marketing of seed potatoes. However, the main buyers of the potato seeds were NGOs (such as World Vision) and government agencies (particularly from the Ministry of Agriculture). Farmers reported that the demand for improved potato seeds from the private sector (traders) had been minimal. NGOs and state agencies distributed seed potatoes to ware potato farmers across the country.

Members of the cooperatives grew seed potatoes on a relatively large scale, average production per grower was 25 tons, and the maximum amount produced by a farmer in the study period was 170 tons. However, the amount of seed potatoes sold via the cooperatives was, on average, 4.6 tons. The cooperatives had difficulties to find buyers for the seed potatoes produced by their members. Thus, growers had to sell the remaining seeds as ware potatoes in the local market, usually at a lower price. In the improved seed market, we observed that the private sector hardly played any role. Growers' reliance on NGOs and government agencies for selling their seeds is an indication of the vulnerability of the improved potato seed industry. We observed major volatility in the buying behaviour of these institutional buyers, resulting in high uncertainty for the growers.

During and before 2008/2009, members of the three seed potato cooperatives reported to have sold their seed potatoes to NGOs and government agencies at reasonable prices (up to 600 Ethiopian Birr per 100 kg). These relatively high prices induced incumbent and new seed growers to increase the production of seed potatoes. Subsequently, the number of farmers participating in the specialised seed potato market had significantly increased. For example, in Jeldu district, the number of seed potato cooperatives grew from one to four, with a total supply of about 12,500 tons. In addition, existing seed cooperatives increased their production level. For example, Derara Improved Seed Potato Growers' Cooperative in Jeldu district had doubled its seed production from 1,250 tons in 2008/2009 to 2,600 tons in 2009/2010. However, in 2009/2010 there was hardly any demand for seed potatoes from the institutional buyers. During our visit in April 2010, the two seed potato cooperatives from Welmera district did not receive any purchase request from the usual institutional buyers. Similarly, the seed cooperative in Jeldu could only sell 480 tons out of the total 2,600 tons of improved seed potatoes in storage. We observed the frustration of farmers who had been waiting for a buyer for seven months, keeping their seed potatoes in diffused light stores. A farmer who had still 180 tons of seed potatoes in storage described his frustration as follows:

> Last year, I only produced a small amount of seed potatoes and was able to sell all of them. I was a bit disappointed last season that I grew only on a small plot of land. This season, I decided to use all my land for growing seed potatoes and I did that. Now, it is April. I have been waiting for a buyer since December. If I don't find a buyer in the next couple of weeks, I will lose everything that I was hoping for all the season. I am frustrated.

I don't know what I need to do. Please help us! Inform our desperate situation to the media, to the government that we are losing everything.

This example shows that the specialised seed potato growers are poorly connected to the ware potato growers. Nonetheless, our case study analysis does not necessarily imply that there was excess supply of improved seed potatoes in Ethiopia. It simply highlights the presence of poor coordination between the specialised seed sector and the ware potato sectors in the country.

6.5.2 Survey

From the original sample of 350, questionnaires from 4 farmers could not be used. Thus, our sample for analysis consists of 346 farmers. Table 6.1 summarises the characteristics of the ware potato farmers sample. The average age was 37, the average years of school education was 6, average land holding was 1.5 ha, and the average animal holdings in number of livestock units was 8. The average family size was 10, with a dependency ratio of 1.3, implying a high number of economically inactive members in a household. Over 33% of the respondents had two or more wives. Farmers who practiced polygamy were older (42 years) and less educated (4 school years), and had a larger family size (14). In terms of access to information technology, 67 and 65% of the ware potato farmers had a mobile phone and radio, respectively. As can be observed in Table 6.1, potato contributed to almost 50% of the total household income in the 2009/2010 production season.

Table 6.2 presents the description of chain actors at the central market, Addis Ababa. In the 2009/2010 transaction season, the mean purchased volume was 957 tons for stationed wholesalers, 42 tons for retailers, and 7 tons for hotels. Although stationed wholesalers did not have a direct transaction with potato farmers, they knew the region where the potatoes came from; they stated that 65% of the potatoes were bought from the Shashemene area, our study region. However, the hotel managers did not know the production region; they buy directly from stationed wholesalers based on a prearranged contract. This means that although production region or variety names are used to signal quality to traders, this signal becomes less relevant for final customers, such as hotels and restaurants. Thus, traders play an important role in translating the final customers' quality preferences into demand for specific varieties or potatoes from specific regions.

6.5.3 Farmer-based characterisation of improved and local varieties

Characterising the quality attributes of local and improved varieties using farmers' knowledge can provide better insights into the quality differences among currently used varieties in general and between the main improved and local varieties in particular. The research question addressed in this part was 'How do ware potato farmers characterise currently used varieties in terms of production- and market-related quality attributes?'

Table 6.1. Characteristics of sampled potato farmers (n=346).

Variables	Total
Male headed households (% yes)	96
Marital status(% yes)	
Unmarried	3.5
Widow	0.3
Married (only one wife)	63
Married (>=2 wives)	33.2
Other demographic characteristics	
Age (years)	36.8
Family size	9.6
Education of the respondent (years in school)	5.7
Highest education in the family (years in school)	8.3
Dependency ratio[1]	1.3
Wealth (as approximated by)	
Land (owned) in ha	1.5
Total livestock units TLU[2]	8.1
Access to information	
Presence of a mobile phone (% yes)	66.8
Presence of a radio (% yes)	65
Presence of a TV (% yes)	9.5
Income from potato sales (2009/2010)	
Mean value (in birr)[3]	9,905
% of potato income from total income	49

[1] Measures the ratio between dependents and labour force within the family.
[2] TLU = tropical livestock units (250 kg), used as a common unit to describe livestock numbers of various species as a single figure: oxen/cow = 1 TLU; calf = 0.25 TLU; heifer = 0.75 TLU; sheep/goat = 0.13 TLU; young sheep/goat = 0.06 TLU; donkey = 0.7 TLU.
[3] 1 US$ ~12.60 birr.

Table 6.2. Potato buyers' knowledge of production region (buyers at the central market Addis Ababa).

Mean values	Hotels (n=11)	Stationed wholesalers (n=10)	Retailers (n=13)
Amount purchased in 2009/2010 (in tons)	6.8	956.8	42.1
Knows production region of potatoes purchased (% yes)	0	100	15.4
If yes, potatoes purchased originated from (in %)			
West Arsi Zone (the study area)	0	65	70
East Arsi zone (South East)	0	28	26.5
Holeta area (West)	0	11.7	7
Gojam (North West)	0	5.3	0

Farmers' assessment of production-related quality attributes of the local and improved varieties is reported in Table 6.3. The improved varieties, with the exception of Bule, scored higher than the local varieties as to yield. Likewise, the improved varieties scored the highest in terms of disease tolerance. Regarding drought tolerance, the most common local varieties were assessed similar to the improved varieties. However, differences exist concerning maturity period. While Gudane and Jalene require on average 123 days to mature, Nech Abeba, the dominant local variety, matures on average in 101 days. As to the intensity of crop management, the local varieties Nech Abeba and Aga Zer scored higher than the improved varieties Gudane and Jalene, implying that the local varieties require more time for land preparation, planting and weed control than the improved varieties. New varieties tend to be more demanding in terms of crop management than local ones, which is not the case in our results. This might be because the farmers in our survey only grew the improved varieties on a small scale; they would not recognise the extent of crop management practices needed to grow improved varieties as they would do for local varieties, which were grown on a relatively large scale.

Table 6.4 presents farmers' assessment of market-related quality attributes. The mean scores for cooking quality and taste appear to be similar; Agar Zer, Bule, and Jalene scored the highest in terms of cooking quality and taste. All the improved varieties scored the highest in storability, while the local variety Nech Abeba scored the highest in stew quality. However, although the improved varieties could be stored for longer periods than the local varieties, ware potato farmers generally do not keep potatoes after harvest because of lack of modern storage facilities and the high risk of quality losses in the traditional storage facilities. In terms of tuber size, Jalene and Gudane, followed by Nech Abeba, scored the highest. In terms of colour of appearance, Bule and Key Dinch are red, while the others are white/yellowish. As to shape, also a dichotomy can be found, with Agar Zer and Key Dinch having an oval shape, while

Table 6.3. Farmer based characterisation of improved and local potato varieties: production related quality attributes (mean scores).

Variety	Yield (100 kg per Timad)	Maturity period (days)	Tolerance to disease (scale 1-5)	Tolerance to drought (scale 1-5)	Management intensity (scale 1-5)
Aga Zer	26.5 (10.5)	88 (10)	3.7 (1.2)	3.8 (1.1)	3.7 (1.0)
Nech Abeba	31.2 (11.8)	101 (15)	3.5 (1.2)	3.9 (1.2)	4.3 (0.9)
Key Dinch	21.7 (8.3)	70 (11)	2.6 (1.2)	2.8 (1.2)	2.9 (1.1)
Key Abeba	21.5 (7.8)	82 (15)	1.8 (0.9)	2.2 (1.1)	3.1 (1.2)
Gudane[1]	43.8 (18.3)	122 (16)	4.1 (1.0)	3.8 (1.2)	3.7 (1.3)
Jalene[1]	45.7 (20.1)	123 (18)	4.2 (1.2)	3.7 (1.3)	3.7 (1.3)
Bule[1]	24.4 (11.9)	112 (20)	3.8 (1.1)	3.5 (1.4)	2.8 (1.3)

[1] Improved varieties; standard deviations are given in brackets.

the others having a round or semi-round shape. Because the scores for taste and cooking quality were similar, and storability was not considered important, taste and storability were dropped from further analysis.

6.5.4 Comparison of improved and local varieties

We statistically tested the difference in production-related quality attributes within the varieties using Friedman Test, which is commonly used in the literature to compare overall mean ranks of ordinal data (Friedman, 1937). However, we only focused on Nech Abeba and Agar Zer (local varieties) and Gudane and Jalene (improved varieties) as these varieties were the most commonly grown by ware potato farmers. Among the four varieties, the Friedman Test results suggest (see Appendix 6.1) the presence of overall significant differences related to the mean ranks of yield, maturity period, and disease tolerance, but no significant differences regarding drought tolerance and crop management intensity. That is, farmers do not perceive that these varieties vary as to drought tolerance and intensity of crop management, while they perceive a difference in yield, maturity period, and disease tolerance. In terms of market-related quality attributes, the Friedman Test result also shows (see Appendix 6.2) that there is an overall significant difference between the mean ranks of tuber size, stew quality, and cooking quality.

The results in Appendix 6.1 and 6.2 show the presence of an overall significant difference in the mean ranks of production- and market-related quality attributes. As the Friedman Test results do not tell which varieties differ, we were also interested to know the quality difference among the varieties. A post-hoc test can determine whether significant differences exist between pairs of the different varieties (Sheldon *et al.*, 1996). Subsequently, we run a multiple comparison using the Wilcoxon Signed-Rank Test. Accordingly, the paired results between Agar Zer and Nech Abeba, Agar

Table 6.4. Farmer based characterisation of improved and local potato varieties: market-related quality attributes (mean scores).

Variety	Tuber size (scale 1-5)	Stew quality (scale 1-5)	Cooking quality (scale 1-5)	Storability (weeks)	Taste (scale 1-5)	Colour
Aga Zer	3.4 (0.7)	3.8 (1.0)	4.6 (0.8)	11.5 (5.5)	4.7 (0.6)	white
Nech Abeba	3.9 (0.3)	4.8 (0.6)	3.9 (0.9)	8.7 (5.3)	3.9 (0.9)	white
Key Dinch	2.3 (0.9)	2.5 (1.1)	3.1 (1.1)	11.1 (5.6)	3.2 (1.1)	red
Key Abeba	2.2 (1.1)	2.6 (1.1)	2.2 (1.0)	8.5 (4.5)	2.0 (1.0)	white
Gudane[1]	4.4 (0.6)	3.3 (1.0)	3.5 (1.1)	13.0 (7.6)	3.3 (1.0)	white
Jalene[1]	4.8 (0.4)	3.4 (1.1)	3.9 (1.1)	11.9 (6.6)	4.0 (0.9)	white
Bule[1]	3.0(1.0)	2.2 (1.3)	3.8 (1.4)	15.6 (7.6)	4.2 (1.1)	red

[1] Improved varieties; standard deviations are given in brackets.

Zer and Jalene, and Nech Abeba and Gudane are not significantly different regarding disease tolerance, stew quality, and cooking quality, respectively. Furthermore, the two improved varieties, Gudane and Jalene, do not significantly differ in terms of yield and maturity period. In all the remaining combinations, the results are highly significant (Table 6.5). The results show that, with the exception of cooking quality related to Nech Abeba and Gudane, the improved and local varieties are significantly different in terms of the main production- and market-related quality attributes.

In sum, the results of Table 6.3, 6.4, and 6.5 confirm that ware potato farmers indeed see quality differences among varieties, particularly between the improved and local varieties.

6.5.5 Quality attributes influencing variety choice

Having characterised the quality attributes of the local and improved varieties, we now turn to address the specific question 'Which quality attributes influence ware potato growers' choice for specific varieties and how does this choice relate to the preferences of downstream actors?'

As can be observed in Table 6.6, 80% of farmers grew the local variety Nech Abeba. It appears that farmers who grew Nech Abeba were largely motivated by the high market demand, which accounted for 57%. This implies that farmers' decision to grow a specific variety is influenced by the market demand more than the price. This makes sense because price is largely determined by the supply and demand conditions and is uncertain at the time farmers make the decision to grow a specific variety. The second most grown variety was the local variety Agar Zer (48%). The main reason for growing this variety appeared to be its good cooking quality. It is also interesting to see

Table 6.5. Comparison of production- and market-related quality attributes across varieties (Wilcoxon Signed Ranks Test).

Variety[1]	Tuber size		Stew quality		Cooking quality		Yield		Days to mature		Disease tolerance	
	Z-score	Sign.	Z-score	Sign.	Z-score	Sign.	Z-score	Sign.	Z-score	Sign.	Z-score	Sign.
AZ & NA	-9.26	0.000	-8.81	0.000	-9.82	0.000	-5.56	0.000	-11.48	0.000	-0.75	0.451
AZ & GD[2]	-7.94	0.000	-2.66	0.008	-4.90	0.000	-4.29	0.000	-7.46	0.000	-4.77	0.000
AZ & JL[2]	-11.38	0.000	-2.55	0.011	-3.93	0.000	-5.84	0.000	-9.75	0.000	-4.41	0.000
NA & GD[2]	-6.71	0.000	-5.64	0.000	-1.05	0.293	-4.25	0.000	-6.27	0.000	-3.40	0.001
NA & JL[2]	-11.37	0.000	-7.80	0.000	-2.85	0.004	-6.22	0.000	-8.71	0.000	-4.35	0.000
GD & JL	-4.74	0.000	-2.75	0.006	-2.93	0.003	-0.51	0.609	-0.86	0.389	-2.70	0.007

[1] Local varieties: AZ = Aga Zer; NA = Nech Abeba. Improved varieties: GD = Gudane; JL = Jalene.
[2] Comparisons between local varieties and improved varieties.

that production-related quality attributes have received low importance in farmers' variety choice. Of the farmers who grew Nech Abeba, only 20% stated that they were motivated by its high yield. Likewise, 14% of farmers who grew Agar Zer claimed that early maturity was the main reason for growing this variety. Of the local varieties, Key Dinch, which was grown by 21% of the respondents, appeared to have been chosen for its production-related attributes; 73% of farmers stated that early maturity was the main factor in their decision to grow this variety. Variety Key Dinch seemed to have been used as a 'hunger breaker' because of its shorter maturity period compared to the other potato varieties.

Key Abeba and Bule were the least preferred varieties. This confirms the results displayed in Tables 6.3 and 6.4. Farmers assessed that Key Abeba is highly susceptible to disease and has low stew and cooking quality, while variety Bule has a long maturity period and the lowest stew quality. As expected, varieties Jalene and Gudane appeared to have been preferred because of their better yield and disease tolerance characteristics. However, not many farmers grew these varieties in the 2009/2010 production season. The results suggest that farmers' decision to grow the local varieties was mainly motivated by market-related quality attributes while their decision to grow the improved varieties was largely based on production-related quality attributes (Table 6.6).

Table 6.7 provides an overview of downstream actors' preferences for market-related quality attributes. We only focused on the preferences of downstream actors in the central market, Addis Ababa, because 65% of the ware potatoes available in Addis Ababa were supplied by the collecting wholesalers in our study area. Stationed

Table 6.6. Factors influencing ware potato growers' variety choice.

Variety	% farmers who grew (n=346)	Reasons for growing a specific variety in the 2009/2010 production cycle (in %)						
		High market demand	High price	Good cooking quality	Storability	High yield	Early maturity	Disease tolerance
Aga Zer	48.3	20.4	18	34	7.2	5.4	14.4	0.6
Nech Abeba	79.5	57.1	17.5	1.7	2.2	20.4	1.1	0
Key Dinch	20.5	9.9	2.8	9.9	0	4.2	73.2	0
Key Abeba	3.8	46.2	20.4	5.3	7.7	7.7	12.7	0
Gudane[1]	9	9.7	6.5	12.9	3.2	51.6	0	16.1
Jalene[1]	19.7	2.9	4.4	1.5	4.4	72.1	0	14.7
Bule[1]	1.2	0	0	100	0	0	0	0

[1] Improved varieties.

Table 6.7. Downstream actors' preferences for market-related quality attributes (mean scores from 100 points).

Quality attributes	Total (n=34)	Hotel (n=11)	Stationed wholesaler (n=10)	Retailer (n=13)
Colour	37.3	35.4	30.3	44.2
Tuber size	32.8	32.6	38.2	28.8
Storability	7.2	5.0	11.5	5.8
Tuber shape	6.9	15.8	6.0	0
Stew quality	6.2	6.8	6.7	5.4
Price	5.8	2.1	2.0	11.9
Cooking quality	4.1	2.3	5.3	4.6

wholesalers had high priority for tuber size, followed by tuber colour and storability. Retailers mostly preferred tuber colour, followed by tuber size. For hotels, tuber colour received the highest weight, followed by tuber size and shape. Generally, cooking quality, price, stew quality, and storability were assigned low scores by traders. As to stew quality, this result is surprising because most urban households consume potatoes in stew. One possible explanation is that stew quality characteristics might have been captured in tuber size. Generally, tuber size and colour are the most important quality attributes influencing downstream actors' preferences for specific varieties.

With regard to tuber size, 80% of wholesalers, 46% of retailers and 36% of hotel managers preferred large tubers. Generally, large tuber size is preferred to small, medium, and very large tuber size (Table 6.8). In terms of colour, there is not much difference between the most common local and improved varieties. However, with tuber size, the two main improved varieties appeared to have very large tuber size compared to the local varieties (Table 6.4).

Taking the above results together, we can better explain why certain varieties are more preferred than others. It appears that the local variety Nech Abeba was more popular among both farmers and downstream actors compared to the improved varieties currently available. Improved varieties Jalene and Gudane appeared to have a very large tuber size, which is not what trader prefer. According to the stationed wholesalers and retailers in Addis Ababa, very large tubers tend to have lower quality when used in stew, which is an important quality attribute for household consumers. Furthermore, household consumers in general buy in small quantities from retailers, and very large tubers could create measurement problems.

Even though ware potato farmers attempted to align their quality preferences with those of downstream actors by growing the preferred local variety Nech Abeba, this cannot reduce quality concerns arising from poor crop management, storage

Table 6.8. Desired tuber size by downstream actors (%).

	Hotel (n=11)	Stationed wholesaler (n=10)	Retailer (n=13)
very large	27.3	20.0	15.4
large	36.4	80.0	46.2
medium	36.4	0	15.4
small	0	0	23.1

condition, and transportation. Stationed wholesalers at the central market, Addis Ababa, reported a number of problems – premature tubers, spoilage losses, blackened tubers, and physical damage. We also observed these quality problems when we visited the potato market in Addis Ababa. Traders have responded to these quality concerns by applying various coordination mechanisms. For example, collecting wholesalers attempted to monitor the quality of potatoes by visiting specific ware potato farms and by involving themselves in the harvest and transport activities. For the 2009/2010 production season, 56% of the farmers reported that harvesting was carried out by the collecting wholesalers. The dyadic relationship between collecting wholesalers and stationed wholesalers was based on trust, where close personal ties and repetitive interactions appeared to have played an important role in governing the relationship (Table 6.9). Accordingly, 60% of the stationed wholesalers had their own preferred suppliers among the collecting wholesalers. Also, 82% of the hotels and 92% of the retailers in the central market, Addis Ababa, had their own preferred suppliers among stationed wholesalers. Written contracts were only used to govern the transactions between big hotels and stationed wholesalers (Table 6.9).

6.6 Conclusions

Table 6.9. Coordination mechanisms used for the transaction between downstream actors.

Type of coordination mechanism	Hotel (n=11)	Stationed wholesalers (n=10)	Retailers (n=13)
Preferred supplier (% yes)	81.8	60	92.3
Written contract (% yes)	33.3	0	0

Using a case study on the specialised seed potato growers and a survey among ware potato growers and downstream actors, the study has provided insights in the alignment of the quality preferences of different actors along the Ethiopian potato supply chain. To better understand which quality attributes influence variety choices, the paper has analysed the link between the specialised seed potato growers and the ware potato growers, has characterised different potato varieties, and has compared local and improved potato varieties.

The case analysis showed that improved potato varieties Jalene and Gudane were widely grown by the specialised seed growers in the highlands. Despite claims of insufficient supply of potato seed of improved varieties (Hirpa *et al.*, 2010), no evidence of insufficient supply was found during the study period. However, these varieties were not very popular among ware potato farmers in the study area. Common explanations for low demand among ware potato growers are a lack of information on how to grow improved varieties and a lack of information on the availability of improved varieties. While ware growers themselves did not buy tubers of these improved varieties, most of the demand comes from institutional buyers like NGOs and government agencies (which distributed the tubers among ware growers in other regions). This demand from institutional buyers, however, turned out to be erratic, leading to high frustration among growers of potato seeds. On the basis of the case study results we can conclude that coordination between two essential parts of the potato supply chain – between seed potato growers and ware potato growers – is poorly developed. This brings us to the second major question of this article: why do ware potato growers choose local varieties above improved varieties? This question entailed the analysis of the extent of alignment of preferences between ware potato growers and downstream actors in the supply chain. But first we had to characterise the main quality attributes of the different varieties, especially distinguishing between production-related and market-related quality attributes.

Seven main varieties were identified and characterised using farmers' knowledge. The aim of such characterisation was to systematically capture the most important quality attributes each variety possesses. Obviously, variety as such is not that important as it is usually short-lived. What matters most are the main quality attributes, which are more stable and determine variety choice. Breeding centres use information on quality attributes to develop new varieties that could satisfy the requirements of ware potato growers and downstream actors.

Farmers' characterisation of the different potato varieties show that there are significant quality differences between the improved and local varieties, particularly with respect tuber size and stew quality (market-related attributes) and yield, maturity period and disease tolerance (production-related attributes). Further analysis shows that the improved varieties are more preferred in terms of their production-related quality attributes while the local varieties are more preferred regarding their market-related quality attributes. We found that ware potato farmers continue to grow local varieties as they consider these having better market-related quality attributes. As a result, demand for improved varieties is low. This result is in line with other literature on commercialisation and variety choice (Asfaw *et al.*, 2012). Commercialisation implies that variety choice is influenced by market requirements.

Among Ethiopian breeding institutes, however, the focus of variety development has always been on yield, disease tolerance, and agro-ecology adaptability to overcome food insecurity challenges in the country (Gebremedhin *et al.*, 2008). Although production-related quality attributes are still important criteria in the selection of specific varieties by the ware potato growers, if varieties do not also possess the quality attributes considered important by downstream actors, ware growers will not choose them. Our results show that even though improved varieties have better production-related quality attributes, ware potato farmers' appeared to have opted for the local varieties due to their superior market-related quality attributes. Thus, ware potato growers continued to grow the local variety Nech Abeba in order to satisfy the quality demanded by downstream actors. In other words, the quality preferences of the ware potato growers and those of the downstream actors are well aligned, while the quality preferences of the breeding institutes + seed potato growers are not well aligned with the preferences of the other actors in the potato chain.

Finally, potato has become an important source of cash for smallholders in Ethiopia. In the case examined, potato contributed about 50% of total household income in the 2009/2010 production season. Thus, research on variety development needs to take into account the quality preferences of downstream actors to enhance the uptake of improved varieties by ware potato farmers. Varieties with better market-related quality attributes would create better market opportunities for the specialised seed potato growers and enhance the uptake of modern potato seeds by the ware potato growers in the Rift valley region, which contributes at least 65% of the ware potato supply to the capital, Addis Ababa.

Acknowledgements

We gratefully acknowledge the Wageningen University International Research and Education Fund (INREF) and the Netherlands Fellowship Program (NFP) of NUFFIC for funding this research.

References

Asfaw, S., Shiferaw, B., Simtowe, F., and Lipper, L., 2012. Impact of modern agricultural technologies on smallholder welfare: evidence from Tanzania and Ethiopia. Food Policy 37: 283-295.
Bijman, J. and Wollni, M., 2009. Producer organizations and vertical coordination. An economic organization theory perspective. In: Rosner, H.J. and Schulz-Nieswandt, M. (eds.) Beiträge der genossenschaftlichen selbsthilfe zurwirtschaftlichen und sozialen entwicklung [Contributions of the cooperative self-help to economic and social development]. LIT Verlag, Berlin, Germany, pp. 231-252.
Cavatassi, R., Lipper, L. and Narloch, U., 2011. Modern variety adoption and risk management in drought prone areas: insights from the sorghum farmers of eastern Ethiopia. Agricultural Economics 42: 279-292.

Emana, B. and Nigussie, M., 2011. Potato value chain analysis and development in Ethiopia. International Potato Center (CIP-Ethiopia), Addis Ababa, Ethiopia.

Friedman, M., 1937. The use of ranks to avoid the assumption of normality implicit in the analysis of variance. Journal of the American Statistical Association 32: 675-701.

Gebremedhin, W., Endale, G. and Lemaga, B., 2008. Potato variety development. In: Gebremedhin, W., Endale, G. and Lemaga, B. (eds.) Root and tuber crops: the untapped resources. Ethiopian Institute of Agricultural Research, Addis Ababa, Ethiopia, pp. 15-32.

Gildemacher, P.R., Demo, P., Barker, I., Kaguongo, W., Woldegiorgis, G., Wagoire, W.W., Wakahiu, M., Leeuwis, C. and Struik, P.C., 2009a. A description of seed potato systems in Kenya, Uganda and Ethiopia. American Journal of Potato Research 86: 373-382.

Gildemacher, P.R., Kaguongo, W., Ortiz, O., Tesfaye, A., Woldegiorgis, G., Wagoire, W.W., Kakuhenzire, R., Kinyae, P.M., Nyongesa, M., Struik, P.C. and Leeuwis, C., 2009b. Improving potato production in Kenya, Uganda and Ethiopia: a system diagnosis. Potato Research 52: 173-205.

Henson, S. and Jaffee, S., 2008. Understanding developing country strategic responses to the enhancement of food safety standards. World Economy 31: 548-568.

Hirpa, A., Meuwissen, M.P.M., Tesfaye, A., Lommen, W.J.M., Lansink, A.O., Tsegaye, A. and Struik, P.C., 2010. Analysis of seed potato systems in Ethiopia. American Journal of Potato Research 87: 537-552.

Hirpa, A., Meuwissen, M.P.M., Van der Lans, I.A., Lommen, W.J.M., Oude Lansink, A.G.J.M., Tsegaye, A. and Struik, P.C., 2012. Farmers' opinion on seed potato management attributes in Ethiopia: a conjoint analysis. Agronomy Journal 104: 1413-1424.

Howard, H., 1974. Factors influencing the quality of ware potatoes. The genotype. Potato Research 17: 490-511.

Jemison, J.M., Sexton, P. and Camire, M.E., 2008. Factors influencing consumer preference of fresh potato varieties in Maine. American Journal of Potato Research 85: 140-149.

Lancaster, K.J., 1966. A new approach to consumer theory. Journal of Political Economy 74: 132-157.

Malone, T.W. and Crowston, K., 1994. The interdisciplinary study of coordination. ACM Computing Surveys 26: 87-119.

Narrod, C., Roy, D., Okello, J., Avendaño, B., Rich, K. and Thorat, A., 2009. Public-private partnerships and collective action in high value fruit and vegetable supply chains. Food Policy 34: 8-15.

Ortiz, O., Orrego, R., Pradel, W., Gildemacher, P., Castillo, R., Otiniano, R., Gabriel, J., Vallejo, J., Torres, O. and Woldegiorgis, G., 2013. Insights into potato innovation systems in Bolivia, Ethiopia, Peru and Uganda. Agricultural Systems 114: 73-83.

Pretty, J.N., Morison, J.I. and Hine, R.E., 2003. Reducing food poverty by increasing agricultural sustainability in developing countries. Agriculture, Ecosystems and Environment 95: 217-234.

Ruben, R., Van Boekel, M., Van Tilburg, A. and Trienekens, J., 2007. Tropical food chains: governance regimes for quality management. Wageningen Academic Publishers, Wageningen, the Netherlands.

Schulte-Geldermann, E., Gildemacher, P. and Struik, P., 2012. Improving seed health and seed performance by positive selection in three Kenyan potato varieties. American Journal of Potato Research 89: 429-437.

Sheldon, M.R., Fillyaw, M.J. and Thompson, W.D., 1996. The use and interpretation of the Friedman test in the analysis of ordinal scale data in repeated measures designs. Physiotherapy Research International 1: 221-228.

Stewart, H., Blisard, N., Bhuyan, S. and Nayga Jr, R.M., 2004. The demand for food away from home: full service or fast food. USDA-Economic Research Service Report No. 829. USDA-Economic Research Service, Washington, DC, USA.

Swinnen, J.F.M. and Maertens, M., 2007. Globalization, privatization, and vertical coordination in food value chains in developing and transition countries. Agricultural Economics 37: 89-102.

Tefera, T.L., Sehai, E. and Hoekstra, D., 2011. Status and capacity of farmer training centers (FTCs) in the improving productivity and market success (IPMS) pilot learning woredas (PLWs). International Livestock Research Institute (ILRI), Addis Ababa, Ethiopia.

Appendix 6.1. An overall difference in mean ranks test of the production-related quality attributes across main varieties.

(Friedman Test)

Variety	Yield	Disease tolerance	Drought tolerance	Maturity	Management practices
Aga Zer	1.73	1.8	2.29	1.19	2.27
Nech Abeba	1.83	2.29	2.58	2.13	2.62
Gudane[1]	3.25	2.77	2.61	3.31	2.56
Jalene[1]	3.19	3.13	2.53	3.37	2.55
N	24	63	56	67	56
χ^2	34.5	43.6	2.6	157.6	3
df	3	3	3	3	3
Sig	0.00	0.00	0.464	0.00	0.391

[1] Improved varieties.

Appendix 6.2. An overall difference in mean ranks test of the market-related quality attributes across main varieties.

(Friedman Test)

Variety	Tuber size	Stew quality	Cooking quality	Storability	Taste
Aga Zer	1.37	2.41	3.18	2.6	3.44
Nech Abeba	2.08	3.49	2.26	1.82	2.14
Gudane[1]	2.96	1.9	2.02	2.71	2.01
Jalene[1]	3.58	2.2	2.54	2.88	2.41
N	98	58	71	41	75
χ^2	200.7	61.4	38.7	19.6	69.6
df	3	3	3	3	3
Sig	0.00	0.00	0.00	0.00	0.00

[1] Improved varieties.

7. Keeping up with rising quality demands? New institutional arrangements, upgrading and market access in the South African citrus industry

V. Bitzer[1], A. Obi[2] and P. Ndou[2]

[1]*Royal Tropical Institute (KIT), P.O. Box 95001, 1090 HA Amsterdam, the Netherlands;* [2]*University of Fort Hare, Department of Agricultural Economics & Extension, Private Bag X1314, Alice 5700, South Africa; v.bitzer@gmail.com*

Abstract

The shift towards the use of private quality standards in global agrifood chains has raised concerns worldwide that small-scale farmers become excluded from lucrative export markets. In South Africa, given the historical exclusion of small-scale farmers from export-oriented agriculture, the government has therefore introduced different new institutional arrangements (IAs) between small-scale farmers and established agribusinesses to promote access to such markets. This chapter aims to analyse these IAs to understand whether and how these IAs contribute to enhanced market access for small-scale farmers. Based on a conceptual framework on quality specifications and upgrading grounded in Global Value Chain analysis, the chapter first discusses the quality demands and standards in the South African citrus sector which manifest in a 'Ladder of Market Access'. The following analysis reveals that IAs are able to promote the required product and process upgrading to include small-scale farmers into global export markets. Further upgrading opportunities, however, remain elusive as agribusinesses manage to position themselves as 'gatekeepers' which places barriers to farmers' involvement beyond the farm gate. These insights provide the basis for a set of practice-oriented recommendations specifically addressing policy-makers and other actors in the South African citrus industry to improve the design of smallholder support programmes.

Keywords: South African citrus industry, institutional arrangements, small-scale farmers, quality, upgrading; market access

7.1 Introduction

High value markets for fresh and processed fruit are increasingly recognised for their potential to contribute to agricultural growth and poverty reduction in developing and transition countries, especially in sub-Saharan Africa (World Bank, 2007). The African country with the longest history of integration into global fresh chains is South

Africa, particularly when it comes to citrus fruit. Exports of citrus, initially exclusively to the UK, started already in the first decades of the last century and had gained a strong foothold in Western European markets by the 1960s (Mather and Greenberg, 2003). Over the past two decades, however, access to such export markets has become conditional on an increasing number of requirements, including conformity to rising quality demands and various private quality standards. Whilst historically, food grades and standards were viewed as public domain issues necessary to address imperfect and asymmetric information leading to market failure (Henson and Reardon, 2005; Reardon et al., 2003), contemporary agrifood chains are increasingly characterised by the proliferation of private quality requirements (Dolan and Humphrey, 2004; Hatanaka et al., 2005; Henson and Reardon, 2005). There is general agreement that this trend reflects the dominance of so-called lead firms in global agrifood chains, notably large retail companies, which are able to set the requirements for upstream producers (Gereffi et al., 2005; Altenburg, 2006). In turn, this emphasis on quality standards is motived by changes in consumption patterns, growing demand for 'safe' or 'ethical' food, competitive struggles among retailers, the devolution of the state in matters of quality control and the resulting privatisation of market governance, among others (Henson and Humphrey, 2009; Henson and Reardon, 2005).

In the South African citrus sector, these developments were paralleled by far-reaching changes in the domestic regulatory environment. From being one of the most protected and regulated sectors of the South African economy, agriculture was deregulated and liberalised in 1997, exposing farmers and agribusiness for the first time to international market forces (Van Dijk and Maspero, 2004). Mather and Greenberg (2003) observe that the two sets of transitions, one at the global level and the other at the national level, have led to a sharpened differentiation 'between growers able to take advantage of the new opportunities [of export chains] and those who are not' (Mather and Greenberg, 2003: 408). For the most part, this differentiation runs along racial lines between mostly white, large commercial producers with access to modern technology and mostly black, small-scale farmers[1] with only rudimentary production technology (Greenberg, 2003). Particularly, the latter face increasing exclusion from lucrative export markets due to smaller production volumes, poorer and inconsistent product quality, lack of resources, and lack of institutional support (Biénabe et al., 2011; Louw et al., 2008).

Similar observations on the exclusion of small-scale farmers have been made for other African supplier countries of fresh fruit products (Dolan and Humphrey,

---

[1] For the purpose of this chapter, the terms 'small-scale' and 'smallholder' farmers are used interchangeably to refer to mostly black, resource-poor, but not necessarily subsistence-oriented farmers. These farmers continue to be marginalised in the mainstream agricultural economy. Chikazunga and Paradza (2012) use the term 'double-barrelled exclusion': initially excluded by the past regimes along racial lines and now excluded by market forces. For this reason, such small-scale farmers are often labelled 'emerging farmers' in the public discourse to indicate the desired process of development towards commercial farming and associated inclusion in mainstream agriculture. Here the more neutral terms 'small-scale' and 'smallholder' farmer will be used to avoid any implicit assumptions about the development trajectory of such farmers.

2004). However, exclusion from global chains does not necessarily imply a permanent situation, as producers who are initially unable to comply with rising quality demands can potentially transform their production practices and receive access to export chains (Perez-Aleman, 2011). In the literature on global value chains (GVCs), these change processes are referred to as 'upgrading', implying changes in the activities of developing country producers to make better products, produce more efficiently or move into more rewarding activities. Stories of successful upgrading are usually explained through the potential to learn from global buyers who, to reduce the risk of supplier failure, may support their suppliers and transfer information on production-related issues (Humphrey and Schmitz, 2002; Saliola and Zanfei, 2009).

However, in many cases global lead firms actually provide rather little support to help suppliers in upgrading their activities (Pietrobelli, 2008). Particularly in fresh fruit chains, retailers have outsourced much of the value chain organisation and have therefore limited direct contact with suppliers in developing countries (Dolan and Humphrey, 2004). This places renewed attention to the role of public policies in creating the institutional conditions necessary to help smallholder farmers upgrade and receive access to global chains. This is especially relevant in South Africa, where post-apartheid policies explicitly aim at uplifting previously disadvantaged smallholder farmers to become commercially viable units by upgrading their productive capacities. Over the years, the government has implemented a range of interventions; often in the context of land reform and Black Economic Empowerment (BEE) policies, including the Black Economic Empowerment in Agriculture (AgriBEE). However, success has been limited and several smallholder farms have collapsed, unable to access the lucrative export markets in the face of continued limited institutional support (Louw et al., 2008; Vermeulen et al., 2008). As a result, the share of smallholder products in formal marketing channels for fresh products remains low (Doyer et al., 2007; Vermeulen et al., 2008).

To address this situation, the South African government has been promoting new types of institutional arrangements (IAs) since the mid-2000s to link farmers to markets, including strategic partnerships, joint ventures and a new generation of farmworker-equity schemes (Department of Rural Development and Land Reform, 2011). These institutional arrangements are based on the active involvement of local agribusinesses, such as processors, marketing and export agents, in assisting small-scale farmers to receive access to mainstream agricultural markets. This is grounded in the assumption that the inclusion of smallholder farmers hinges upon (a) the cooperation among value chain actors, and (b) the provision of extensive support to such farmers (Amekawa, 2009; Henson et al., 2005). For both aspects, local agribusinesses are deemed critical due to their resources, capabilities and positioning along value chains, and thus take a prominent role in the South African context.

Despite the public promotion of IAs in South Africa, especially in the citrus sector, little is known on whether and how these IAs contribute to enhanced market access for smallholder farmers. There is a general understanding that smallholder farmers need to comply with the quality demands required by global buyers and that policy interventions should respond to these quality demands accordingly. Therefore, this chapter first aims to assess what these quality demands are in the citrus sector and how they emerged. This is important for the second aim of the chapter, which is to understand how the new IAs contribute to quality improvement and upgrading to enhance market access for small-scale citrus farmers. Together, this serves to provide recommendations for future policy responses to support smallholder farmers.

The paper is based on a critical review of the literature and uses the global value chain (GVC) approach to understand how the South African state is trying to renegotiate the integration of small-scale farmers into global value chains and what rooms to manoeuvre and to upgrade it creates for these farmers. Within this strand of literature, a vibrant debate on governance, quality and upgrading in value chains has emerged since the late 1990s, which will be used to delineate the conceptual framework of this chapter. The chapter then proceeds to shed light on how quality is defined in practice in the South African citrus sector to better understand what market access actually means in this case in point. The discussion starts by tracing the historic development of the sector from the early 1900s up until the current industry structures, which unfolds around a marked shift from a single export system and volume-based production to a buyer-driven chain and quality-based competition. The implications thereof provide the setting for the analysis of the new institutional arrangements promoted by the South African government, which seeks to unravel the upgrading opportunities for smallholder farmers. The final section reflects on the implications of the IAs within the wider context of post-apartheid South Africa, comparing the upgrading opportunities of commercial versus small-scale farmers. Finally, in the recommendation section, two critical questions offer food for thought for the design of smallholder support programmes.

7.2 Governance, quality and upgrading in global value chains

GVC analysis emerged in the mid-1990s as an analytical approach to examine the complexities of economic development along the chain of actors involved in the production, distribution and consumption of particular goods or services. Scholars have largely focused on the governance of global value chains, referring to the power relationships between actors that determine how financial, material and human resources are allocated along the chain (Gereffi *et al.*, 1994). According to Humphrey and Schmitz (2001), governance entails the setting and enforcing of parameters under which other actors in the chain operate. Initially governance was perceived as a question of producer-driven versus buyer-driven chains, alluding to the location of the power centre of global chains. Agrifood chains have been shown to fall into the category of buyer-driven chains where global buyers hold considerable power despite lacking ownership over most parts of the chain (Dolan and Humphrey, 2000, 2004). Their degree of control is such that they are able to retain the majority of value added and limit the choices and strategies available to producers to increase their gains from participating in global chains (Kaplinsky, 2000).

Since the early 2000s a shift has gradually occurred in GVC research to concentrate on the patterns of inter-firm relationships and coordination (Gereffi *et al.*, 2005). Governance in the sense of coordinating inter-firm relationships entails the transmission of information of relevant production parameters, including product quality, delivery time, process efficiency, environmental, social and labour requirements, which suppliers have to meet in order to participate in global chains (Altenburg, 2006). More complex requirements are often associated with governance forms that feature higher degrees of vertical coordination to ensure that supply matches demand (Maertens and Swinnen, 2009; Reardon *et al.*, 2003). According to Humphrey (2005), compliance with rules about quality is critical to guarantee that firms enjoy sustained access to profitable markets. Hence, from a supplier perspective, the goal of quality improvement is 'to satisfy the expectations of the consumer' (Ruben *et al.*, 2007: 28).

Chain governance in a situation where quality specifications are involved draws heavily on the subjective nature of quality conception. Ponte and Gibbon (2005) therefore claim that governance in chains does not only depend on drivenness or direct coordination, but also on the ability of a given actor to define 'broader narratives about quality circulating within society more generally' (Ponte and Gibbon, 2005: 3). This insight frames the beginning to the most recent theorising on governance in the GVC literature, i.e. governance as a matter of defining and organising a socially constructed specification of quality. The point of departure for this view of governance is the growing importance of different quality attributes in agrifood chains over recent years. As quality came to include not only objectively measurable attributes, but increasingly also subjective elements, researchers began to criticise the neglect of quality specification as a critical aspect of chain governance and key source of power of lead firms (Busch and Bain, 2004; Ponte and Gibbon, 2005).

A broad distinction is often made between intrinsic and extrinsic quality attributes, where the former relate to inherent characteristics of a product, such as physical properties (size, colour, etc.) and the latter pertain to characteristics that are externally added and modifiable (e.g. packaging, brand, price). Petzold *et al.* (2008) note that both product-related attributes, like taste, appearance and nutritional value (real or perceived), and process-related attributes, such as ethical and environmental considerations, play a role in the process of quality conception. Hence, quality can refer to the product itself or to the process and conditions under which the product is produced. Particularly in global food chains, emphasis has shifted towards subjective quality attributes which cannot be objectively measured (credence attributes) and which relate to the process and methods of production (Ponte and Gibbon, 2005), including food safety, ethical and environmental attributes.

Among the strategies that are adopted to meet increasing quality standards, *upgrading* has received considerable attention in the GVC literature and in practice (Humphrey and Schmitz, 2002; Gibbon and Ponte, 2005; Giuliani *et al.*, 2005). Four categories of upgrading are commonly distinguished (Humphrey and Schmitz, 2002): (1) process upgrading (reorganising production processes for greater efficiency and quality); (2) product upgrading (more sophisticated products); (3) functional upgrading (acquiring new functions to increase the overall skills content of activities); and (4) inter-sectoral upgrading (moving into new productive activities). Upgrading opportunities for developing country suppliers are thought to be linked to the governance structures prevailing in global chains, as lead firms can both stimulate certain types of upgrading, e.g. process and product upgrading, by transferring some information and limit other types of upgrading, e.g. functional upgrading, by ensuring continued information asymmetry (Humphrey and Schmitz, 2002; Saliola and Zanfei, 2009).

Upgrading opportunities can also be influenced by targeted public policies (Memedovic, 2008; Pietrobelli, 2008). This has, however, mostly been ignored in the literature, as the GVC framework has generally been weak in integrating local-level institutions into its analysis and policy advice (Selwyn, 2008). Neilson and Pritchard (2009) therefore argue that the concept of upgrading serves as a bridge between chain governance and the local conditions into which producers are embedded. Hence, looking at national strategies and policies that influence upgrading is equally important as stressing the role of lead firms in transferring knowledge along the chain (Memedovic, 2008), especially with regard to development policy formulation (Selwyn, 2008).

Processes of upgrading have generally been associated with increased value added as a prerequisite for enhanced smallholder livelihoods (Giuliani *et al.*, 2005). Particularly functional upgrading into new tasks is often associated in the literature with increased value added (Pietrobelli, 2008). However, scholars also acknowledge that functional upgrading into new tasks requires highly specialised management and entrepreneurial skills and is highly capital intensive. Exploring new opportunities at the same task, i.e. product and process upgrading, may be equally profitable and more easily available to new or small producers in developing countries (Gibbon and Ponte, 2005; Pietrobelli, 2008). Other observers have argued that quality improvement through product and process upgrading may serve to obtain market access, but does not automatically lead to higher value for producers (Ponte and Ewert, 2009; Mather, 2008). Hence, the link between different types of upgrading and enhanced value added is not clear-cut and may vary from case to case. This suggests that – rather than being about increased value capture – upgrading is fundamentally about the question to what extent value chains are inclusive or exclusive and what strategies producers may pursue to go from being excluded to being included. An analysis studying the process of moving from exclusion to inclusion would therefore need to look at (1) where exactly upgrading occurs; (2) how upgrading occurs; (3) what the purpose of upgrading is; and (4) how this influences future upgrading strategies.

7.3 Understanding quality and market access in the South African citrus industry

In this section, the evolution of quality specifications and quality standards in the South African citrus industry is examined alongside the response of the sector to quality trends.

7.3.1 The evolution of quality specifications and quality standards in the South African citrus industry

Over the last two decades, various developments have impacted fundamentally on South Africa's citrus sector (Figure 7.1). Until the mid-1990s, citrus was exported through a single channel exporter (South African Cooperative Citrus Exchange; short SACCE) with uniform quality standards based on public regulation and inspection through a public export control board (Dixie, 1999). This system obscured differences in quality across growing regions and helped in the marketing and promotion of citrus in the UK and Europe (Mather, 1999). To increase its control over production methods, cultivars and varieties produced, SACCE set up an impressive research and extension capacity (Mather and Greenberg, 2003). This provided the foundation for a coordinated strategy to supply lucrative markets with better quality and sought-after varieties and less remunerative markets with poorer quality and standard varieties (Mather, 1999). By virtue of its monopolistic position, SACCE was even able to disregard some quality demands placed by overseas retailers (Gibbon and Ponte, 2005).

However, by the early 1990s, the SACCE's foothold on European markets was challenged on several fronts, including new competition from other Southern Hemisphere producers and international sanctions against South Africa (Dixie, 1999; Mather and Greenberg, 2003). The system established by SACCE proved rigid and

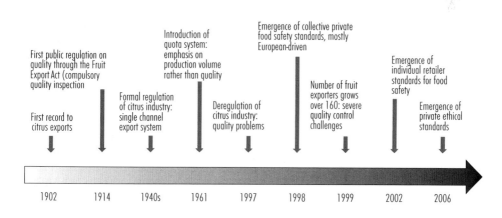

Figure 7.1. The development of 'quality' in the South African citrus industry.

inflexible in the face of new, increasing quality demands by UK and European retailers. As fruit delivered by growers was pooled by variety and size and producers were paid on the basis of volumes and variety (Mather and Greenberg, 2003), high volume rather than high quality production was stimulated (Mather, 1999). To address these challenges and respond to the changing global market, SACCE transformed itself into a private company, Outspan International, in the early 1990s and restructured the pool system. Yet, its strongest days were numbered. In 1997, the South African fruit industry was deregulated and liberalised. The single marketing channel was abolished and a myriad of new exporters entered the market (Mather and Greenberg, 2003).

Quality control became a tremendous challenge in the first years after deregulation. Firstly, vital services to growers such as research support, extension and information dissemination were eliminated. Secondly, the export control board experienced substantial capacity shortages when going overnight from having one main client in citrus to about a couple of hundred clients (De Beer *et al.*, 2003). This compromised its ability to effectively control the quality of citrus exports in the immediate deregulation period (Mather, 2008). Finally, unscrupulous export agents saw deregulation as an opportunity to make money without adhering to quality standards (De Beer *et al.*, 2003). Unable to cope with the new market environment, by 2000 the South African citrus industry found itself in deep crisis, export earnings plummeted and many farms were liquidated (Mather and Greenberg, 2003).

The vacuum created in the turbulences of deregulation was quickly filled by the increasing power of major retailers from consumer countries, transforming the industry from exporter-driven towards a buyer-driven chain, as happened also in other African fresh fruit export chains. This resulted in a shift from volume-based to quality-oriented production. Not trusting the quality standards and assurances of developing countries or not considering them sufficient, retailers started developing their own private quality standards, both for primary production (pre-farm gate) and processing of citrus (post-farm gate). Of particular importance was the development of the EurepGAP (now GlobalGAP) standard in the mid-1990s as a first attempt by large European retailers to harmonise their minimum standards on Good Agricultural Practices (GAPs) and food safety. Shortly afterwards, individual retailers developed their own GAP standards in addition to GlobalGAP, such as Tesco (Nurture, formerly: Nature's Choice) and Marks and Spencer (Field to Fork), leading to a plethora of food safety standards in primary production (Table 7.1).

The sole focus on food safety issues did not last long, however. After repeated scandals about poor working conditions on South African farms, overseas retailers increasingly broadened their quality specifications for suppliers to include social and environmental aspects. Since 2006/2007, several UK retailers have demanded compliance with the base code of the Ethical Trading Initiative (ETI), whereas retailers from continental Europe have started asking for suppliers to be certified against the Business Social Compliance Initiative (BSCI).

Table 7.1. Key private process standards in the citrus industry.

Pre-farm gate				Post-farm gate	
Food safety standards	Demanded by	Social and environmental standards	Demanded by	Pack house standards	Demanded by
GlobalGAP	All UK and EU retailers; some buyers in other export markets; some domestic retailers in SA	Ethical Trading Initiative (ETI) base code	All UK retailers	HACCP (SANS 10330)	All EU and UK retailers
Nurture	Tesco	Business Social Compliance Program (BSCI)	Growing number of EU retailers	British Retailer Consortium (BRC) global standards	All UK retailers
Field to Fork	Marks & Spencer	Fairtrade	Small number of mostly UK retailers as niche market products	GlobalGAP produce handling	Several EU retailers
LEAF Marque	Waitrose and other UK retailers	GRASP (GlobalGAP risk assessment on social practices)	Some EU retailers		
Albert Heijn PPP Protocol	Albert Heijn	Organic	Small number of UK and EU retailers as niche market products		
		Carbon footprint	Tesco		

7.3.2 Implications of the new quality trend for market access and responses by the citrus industry

The rise in private quality standards in South Africa's main export markets, together with the process of deregulation, has had a tremendous impact on the citrus industry. Those who wished to remain competitive had to implement different upgrading strategies to improve quality and enjoy continued access to these export destinations.

Firstly, in terms of process upgrading, compliance with private standards for upgrading on food safety and quality aspects has become wide-spread. After an initial period of slow uptake, there was a flurry of GlobalGAP certification by South African growers during 2003-2004. By 2012, most export growers were certified (Barrientos and Visser, 2012).

Secondly, in terms of product upgrading of fruit varieties, production has shifted from mostly oranges to include high-value citrus varieties, such as grapefruit and soft citrus cultivars (Department of Agriculture, Forestry and Fisheries, 2011). Since citrus fruits are perennial crops, which cannot be easily switched once planted, the demand for new varieties has translated into an increase in plantings and production area. Particularly the areas under soft citrus varieties, grapefruit, lemon and new orange varieties have increased over the past few years (Ndou, 2012). By means of these investments into process and product upgrading, farmers with larger production volumes, more attractive fruit varieties and the required private quality standards have managed to become relatively secure, preferred suppliers for markets in the UK and Europe (Mather, 2008).

However, not all growers have been able to pursue this path of quality upgrading. Smaller farms, particularly black farmers, have been increasingly excluded from lucrative export markets due to smaller production volumes, poorer and inconsistent product quality, lack of resources and knowledge (Mather and Greenberg, 2003). The assistance they had received from the single channel exporter has been withdrawn (Mather, 2008) and has not been replaced by adequate public support, resulting in a shortage of essential services for smallholders (Greenberg, 2010). Unable to meet the stringent demands of foreign customers, these farmers are either moving out of citrus or have found themselves forced to sell to markets with lower quality requirements, which generally translates into lower returns (Mather and Greenberg, 2003). In this way, the new market environment has reinforced the structural duality of agriculture between small-scale and commercial growers, and the associated market segmentation into European/UK supermarkets and 'other' markets.

Three of these 'other' markets can be distinguished in particular. In terms of exports, so-called 'second tier' markets, including Asia, Russia and the Middle East, have much lower quality requirements. Private standards are generally weak, although increasing demand for GlobalGAP has been noted. Similarly, the domestic retail market in South Africa has seen an increase in private quality requirements. A number of domestic retailers have started asking for food safety audits and partially also social audits. The domestic informal market and the processing sector are the only markets where process standards are not an important consideration (Vermeulen et al., 2006). Hence, they absorb citrus fruit of lower and variable product quality based on visual inspection. However, especially the domestic informal market tends to be heavily over-supplied, offering poor returns for producers (Mather, 2008).

The various quality demands posed by the different markets translate into a 'ladder of market access' (Figure 7.2), illustrating the efforts producers need to undertake if they wish to move from local market outlets to supplying European and UK supermarkets.

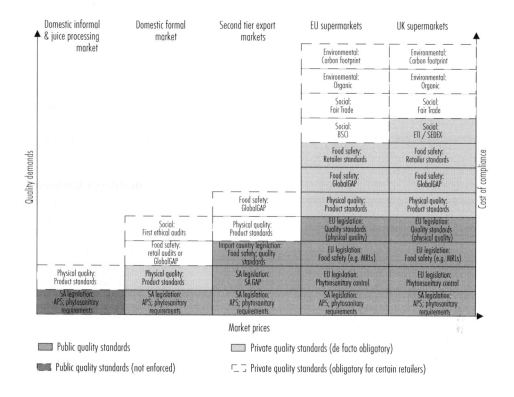

Figure 7.2. The ladder of market access.

7.4 New institutional arrangements for market access of smallholder farmers: what opportunities for upgrading?

Legislative changes have sought to address this dual structure since the democratic transition in 1994 through the implementation of land reform and Black Economic Empowerment (BEE) policies. Initial policies from the mid-1990s until the mid-2000s dealt with the redistribution of farm land to strengthen small-scale growers – with little success. Focussing on asset acquisition rather than skills development, the failure rate of redistribution projects in export-oriented agriculture was high, among others due to the low skills level of many beneficiaries, the lack of post-settlement support, challenges of severe credit constraints and ultimately, inability to meet stringent export market requirements (Greenberg, 2010; Lahiff *et al.*, 2008; Mather and Greenberg, 2003).

More recently, the South African government has sought new ways to ensure that small-scale farmers turn into commercially viable units. In particular, the government has promoted the development of new institutional arrangements (IAs) specifying the collaboration between small-scale farmers (individuals or groups of land reform beneficiaries) and agribusinesses (commercial farmers or marketing agents), facilitated by the state. The expectation is that these new IAs facilitate the transfer of information and the development of skills of small-scale farmers and provide them with market access (Obi and Pote, 2012), while simultaneously recognising the interest of agribusinesses to secure and possibly expand market access (Derman *et al.*, 2010). To ensure that the participants fulfil their responsibilities and commitments, the parties involved sign long-term (5-15 years) contracts and act within a framework set by the state. Three types of IAs are commonly distinguished: strategic partnerships; joint ventures; and farmworker-equity schemes (FES). In strategic partnerships, agribusinesses and small-scale farmers, facilitated by government, agree to jointly manage the farm of the small-scale farmer based on mentorship and service provision, for which agribusinesses receive a management fee. In joint venture partnerships, a new business entity (the operating company) is created in which both the agribusiness and the beneficiaries are shareholders. Benefits (dividends) and risks are allocated according to shares in the joint venture. In FES land reform beneficiaries acquire shares in existing farming operations to farm under the mentorship of established farmers.

Initial research indicates that such IAs are successful in creating increased access to export markets for small-scale producers of citrus fruit by fulfilling three main functions (Bitzer and Bijman, 2014; Lahiff *et al.*, 2012; Louw *et al.*, 2008). Firstly, IAs specify that agribusinesses provide the farmers with information on market developments and market demands in order to adjust activities at the production level. Secondly, IAs ensure that agribusinesses provide production level support to small-scale farmers through training, technical assistance, access to inputs and access to working capital. Thirdly, IAs specify that agribusinesses have exclusive rights to market the products of the small-scale farmers in return for the assistance provided. This ensures that these farmers have access to market infrastructure and that production and marketing activities are coordinated. At the same time, it creates an incentive for agribusinesses to maintain their involvement with small-scale farmers as it gives them the opportunity to increase their export volumes.

Bitzer and Bijman (2014) show how these three functions serve the purpose of meeting the required quality demands of EU and UK retailers through introducing different on-farm changes. *Product upgrading* is achieved by introducing new citrus cultivars in the context of re-planting and land development. *Process upgrading* is achieved by inducing a shift from production practices based on informal standards for physical quality to production practices based on highly formalised, private quality standards and verification by third-party inspection. Hence, virtually all upgrading activities focus on the primary production level, with little attention to *functional upgrading* strategies. On the one hand, such a restricted focus makes practical sense, given the low resource endowment and low skills level of many small-scale farmers. As citrus production is input and capital intensive, maintaining production is already a challenge for these farmers (Lahiff *et al.*, 2012). Meeting international quality requirements presupposes a high level of expertise, which in itself poses high upgrading demands, even if restricted to the production level. Moreover, the orientation towards EU and

UK markets and the high quality demands of these markets also suggests that value adding processes take place from which small-scale producers can benefit.

On the other hand, the lack of functional upgrading poses questions about the future upgrading opportunities for small-scale farmers. As farm net returns are increasingly getting smaller, even for those farmers participating in export markets, producers who functionally upgrade in the value chain appear to be better able to withstand the competitive struggles of the industry than producers focusing only on activities at primary production level (Barrientos and Visser, 2012). Accordingly, there has been an increasing trend over the past decade for producers to export directly to overseas retailers without going through established agribusinesses in order to increase their share of the final consumer price. In other words, export producers have become 'producer-exporters' (Symington, 2008). However, within the context of IAs, such prospects remain blocked for smallholder farmers. Two constraints stand out in particular. The first is a thematic constraint. Whilst smallholders may individually lack the required capacity to export directly, both in terms of skills and production volume, promoting collective action institutions to pool different smallholders may hold potential in this regard (D'Haese et al., 2005). Yet, this is not part of the issues covered by the IAs (Bitzer and Bijman, 2014) and hence, not further explored. The second constraint is structural and shows in the two-fold role of agribusinesses as 'gate-keepers' in IAs. Firstly, they exercise a gate-keeping function with regard to market access, since IA contracts state that agribusinesses provide training to small-scale producers in return for processing and marketing rights. This does not include transparency with regard to market prices and value adding activities further downstream in the value chain, of which producers often are not aware (Fraser, 2007a; Lahiff et al., 2012). Fraser (2007b) cautions that this may enable or induce agribusinesses to take advantage of small-scale farmers. Secondly, agribusinesses act as gatekeepers for knowledge and market information. Any information transfer to small-scale farmers rests on the willingness of agribusinesses to share their knowledge. These agribusinesses enjoy a 'near-monopoly on technical and entrepreneurial skills needed for commercial agriculture' (Fraser, 2007b: 840) in light of the fact that they were the only legitimate commercial entities in agriculture under apartheid. However, this may create a dependency situation for small-scale producers and increase their vulnerability vis-à-vis agribusinesses (Fraser, 2007b). In this way, IAs imitate the quasi-hierarchical chain governance structures that already Humphrey and Schmitz (2001) found to be conducive for product and process upgrading but not for functional upgrading.

7.5 Conclusions

This chapter is set against the background of recent debates about smallholder farmers in sub-Saharan Africa and their growing exclusion from fresh fruit export chains due to rapidly increasing and continuously changing quality demands by international buyers. In South Africa, this debate has an additional dimension and is contextualised in post-apartheid policies, including land reform and AgriBEE, which seek to redress some of the inequalities between mostly black small-scale farmers and largely white large-scale farmers and agribusinesses.

The 'first generation' of government-supported projects to integrate small-scale farmers into the agricultural economy of South Africa had not been very successful, as evident in the collapse of the majority of such farms within a short period of time (Greenberg, 2010). To avoid such failures, in the mid-2000s a 'second generation' of land reform and AgriBEE projects in high-value agriculture was introduced. This time smallholder inclusion was to be stimulated by means of government-facilitated institutional arrangements (IAs) between these farmers and established agribusinesses, especially in export-oriented agriculture, such as the citrus sector.

To contribute to the on-going public debate on whether or not these IAs bring about any positive change for small-scale farmers, this chapter pursued two main objectives. Firstly, it took an in-depth look at what market access in the citrus sector actually means. Secondly, it explored how the new IAs promote upgrading processes for smallholder farmers to facilitate their inclusion into export chains.

As regards the issue of market access, the chapter traced the transformation of the citrus sector from being a single exporter-driven chain emphasising quantity rather than quality to becoming a buyer-driven chain that operates according to the increasing quality demands of overseas retailers. This poses new challenges for South African suppliers. Whereas larger producers mostly managed to address these challenges by means of significant upgrading activities, small-scale producers largely failed to climb the 'Ladder of Market Access' and ended up confined to the domestic informal market.

Concerning the new IAs introduced by the South African government, the chapter showed that on commercial grounds these seem to fare better than the above mentioned first generation. By promoting product and process upgrading to ensure that small-scale farmers meet international quality demands, IAs have successfully included small-scale farmers into global export markets. Hence, they have helped them climb the ladder of market access and enabled them to participate in quality-based competition.

Further upgrading opportunities for small-scale farmers, such as functional upgrading, remain elusive, however. Meanwhile, commercial farmers are busy further integrating into the value chain to become producer-exporters, once again creating a gap between small-scale and commercial farmers – a gap which IAs do not seem to be able to fill. This is grounded in the position of agribusinesses as 'gatekeepers' to knowledge and to market access, which discourages any quality improvement beyond product and process upgrading. Producer involvement does not extend beyond the farm gate, although involving them in cost reduction through enhanced chain efficiency and chain shortening might enable them to strengthen their competitive position vis-à-vis other industry players. Yet, as long as agribusinesses continue to act as double gatekeepers, smallholder farmers are unlikely to develop the required levels of competence for exploring new upgrading pathways.

7.6 Recommendations for practice

In terms of moving forward, the insights from this chapter point towards two main questions which warrant attention from a practitioner perspective. The first question concerns the expansion of upgrading opportunities: How can further upgrading opportunities for small-scale farmers be explored and supported within the context of IAs? The chapter argued that product and process upgrading are generally supported by IAs, while functional upgrading is limited by the current setup of the IAs. Among others, this is due to the non-transparent marketing arrangements which obligate producers to sell to agribusinesses (the packhouses) whilst not knowing the price being paid for their products in export markets or the grade in which the products fall. This calls for new mechanisms to be built into the relationship between smallholders and agribusinesses to ensure chain transparency. New mechanisms may also be needed to ensure a broader, more substantive knowledge transfer from agribusiness to small-scale farmers on issues such as marketing and market channels. This may be another important aspect to increase upgrading opportunities for small-scale farmers, since the type of knowledge transferred affects the type of upgrading available to these farmers (cf. Humphrey and Schmitz, 2002). Finally, it seems rather straightforward that 'smallholder farmers [in South Africa] cannot individually compete against commercial farmers in markets' (Jari *et al.*, 2011: 115). Particularly for functional upgrading, collective action has been shown to work with small-scale farmers in South Africa (D'Haese *et al.*, 2005). The issue of farmer organisation therefore deserves renewed attention. Creating linkages between IAs which operate in geographical proximity may be one way of stimulating the organisation of small-scale farmers involved in these IAs.

The second question concerns the issue of targeted market outlets: Which markets are best suited for small-scale farmers within the context of IAs? The chapter illustrated the different markets and their respective requirements in the 'Ladder of Market Access'. In the context of South Africa, where the domestic informal market – the one which is easiest to access – suffers from oversupply and poor prices, there seems to be a general assumption that small-scale producers need to access EU/UK export markets in order to pursue profitable farming. However, as Figure 7.2 also shows, there are a variety of 'intermediary' markets lying between the opposing ends of 'domestic informal' and 'EU/UK supermarkets'. This suggests the need for a renewed market analysis, ideally on a case-by-case basis, to identify opportunities beyond the traditional market orientation of the South African citrus sector.

References

Altenburg, T., 2006. Governance patterns in value chains and their development impact. European Journal of Development Research 18: 498-521.

Amekawa, Y., 2009. Reflections on the growing influence of good agricultural practices in the global South. Journal of Agricultural and Environmental Ethics 22: 531-557.

Barrientos, S. and Visser, M., 2012. South African horticulture: opportunities and challenges for economic and social upgrading in value chains. University of Manchester, Manchester, UK.

Biénabe, E., Vermeulen, H. and Bramley, C., 2011. The food 'quality turn' in South Africa: An initial exploration of its implications for small-scale farmers' market access. Agrekon 20: 36-52.

Bitzer, V. and Bijman, J., 2014. Old oranges in new boxes? Strategic partnerships between emerging farmers and agribusinesses in South Africa. Journal of Southern African Studies 40: 167-183.

Busch, L. and Bain, C., 2004. New! Improved? The transformation of the global agrifood system. Rural Sociology 69: 321-346.

Chikazunga, D. and Paradza, G., 2012. Can smallholder farmers find a home in South Africa's food system? Lessons from Limpopo province. PLAAS Institute for Poverty, Land and Agrarian Studies, University of the Western Cape, South Africa. Available at: http://tinyurl.com/p6v2mr8.

De Beer, G., Paterson, A. and Olivier, H., 2003. 160 years of export. The history of the perishable products export control board. Perishable Products Export Control Board (PPECB), Cape Town, South Africa.

Department of Agriculture, Forestry and Fisheries, 2011. A profile of the South African citrus market value chain. Pretoria, South Africa.

Department of Rural Development and Land Reform, 2011. Policy framework for the recapitalisation and development programme. Pretoria, South Africa.

Derman, B., Lahiff, E. and Sjaastad, E., 2010. strategic questions about strategic partners: challenges and pitfalls in South Africa's new model of land restitution. In: Walker, C., Bohlin, A., Hall, R. and Kepe, T. (eds.) Land, memory, reconstruction, and justice: perspectives on land claims in South Africa. Ohio University Press, Athens, OH, USA, pp. 306-324.

D'haese, M., Verbeke, W., Van Huylenbroeck, G., Kirsten, J. and D'haese, L., 2005. New institutional arrangements for rural development: the case of local woolgrowers' associations in the Transkei area, South Africa. Journal of Development Studies 41: 1444-1466.

Dixie, G., 1999. Summer citrus: the role and prospects for Southern Africa. In: Jaffee, S. (ed.) Southern African agribusiness: gaining through regional collaboration. World Bank, Washington, DC, USA, pp. 88-138.

Dolan, C. and Humphrey, J., 2000. Governance and trade in fresh vegetables: the impact of UK supermarkets on the African horticulture industry. Journal of Development Studies 37: 147-176.

Dolan, C. and Humphrey, J., 2004. Changing governance patters in the trade in fresh vegetables between Africa and the United Kingdom. Environment and Planning A 36: 491-509.

Doyer, O.T., D'Haese, M., Kirsten, J.F. and Van Rooyen, C.J., 2007. Strategic focus areas and emerging trade arrangements in the South African agricultural industry since the demise of the marketing boards. Agrekon 46: 494-513.

Fraser, A., 2007a. Hybridity emergent: geo-history, learning, and land restitution in South Africa. Geoforum 38: 299-311.

Fraser, A., 2007b. Land reform in South Africa and the colonial present. Social and Cultural Geography 8: 835-851.

Gereffi, G., Humphrey, J. and Sturgeon, T., 2005. The governance of global value chains. Review of International Political Economy 12: 78-104.

Gereffi, G., Korzeniewicz, M. and Korzeniewicz, R., 1994. Introduction: global commodity chains. Commodity chains and global capitalism. Greenwood Press, Westport, CT, USA, pp. 1-14.

Gibbon, P. and Ponte, S., 2005. Trading down: Africa, value chains and the global economy. Temple University Press, Philadelphia, PA, USA.

Giuliani, E., Pietrobelli, C. and Rabellotti, R., 2005. Upgrading in global value chains: lessons from Latin American clusters. World Development 33: 549-573.

Greenberg, S., 2003. Land reform and transition in South Africa. Transformation: critical perspectives on Southern Africa 25: 42-57.

Greenberg, S., 2010. Status report on land and agricultural policy in South Africa, 2010. School of Government, University of the Western Cape, Cape Town, South Africa.

Hatanaka, M., Bain, C. and Busch, L., 2005. Third-party certification in the global agrifood system. Food Policy 30: 354-369.

Henson, S. and Humphrey, J., 2009. The impacts of private food safety standards on the food chain and on public standard-setting processes. Codex Alimentarius Commission, Joint FAO/WHO Food Standards Programme. FAO, Rome, Italy.

Henson, S. and Reardon, T., 2005. Private agri-food standards: implications for food policy and the agri-food system. Food Policy 30: 241-253.

Henson, S., Masakure, O. and Boselie, D., 2005. Private food safety and quality standards for fresh produce exporters: the case of Hortico Agrisystems, Zimbabwe. Food Policy 30: 371-384.

Humphrey, J., 2005. Shaping value chains for development: global value chains in agribusiness. Gesellschaft für Technische Zusammenarbeit, Eschborn, Germany.

Humphrey, J. and Schmitz, H., 2001. Governance in global value chains. IDS Bulletin 32: 19-29.

Humphrey, J. and Schmitz, H., 2002. How does insertion in global value chains affect upgrading in industrial clusters. Regional Studies 36: 1017-1027.

Jari, B., Fraser, G. and Obi, A., 2011. Influence of institutional factors on smallholder farmers' marketing channel choices. In: Obi, A. (ed.) Institutional constraints to small farmer development in Southern Africa. Wageningen Academic Publishers, Wageningen, the Netherlands, pp. 101-120.

Kaplinsky, R., 2000. Globalisation and unequalisation: what can be learned from value chain analysis? Journal of Development Studies 37: 117-146.

Lahiff, E., Davis, N. and Manenzhe, T., 2012. Joint ventures in agriculture: lessons from land reform projects in South Africa. IIED/IFAD/FAO/PLAAS, London, UK.

Lahiff, E., Maluleke, T., Manenzhe, T. and Wegerif, M., 2008. Land redistribution and poverty reduction in South Africa: the livelihood impacts of smallholder agriculture under land reform. University of the Western Cape, Cape Town, South Africa.

Louw, A., Jordaan, D., Ndanga, L. and Kirsten, J.F., 2008. Alternative marketing options for small-scale farmers in the wake of changing agri-food supply chains in South Africa. Agrekon 47: 278-308.

Maertens, M. and Swinnen, J.F.M., 2009. Trade, standards and poverty: evidence from Senegal. World Development 37: 161-178.

Mather, C., 1999. Agro-commodity chains, market power and territory: re-regulating South African citrus exports in the 1990s. Geoforum 30: 61-70.

Mather, C., 2008. The structural and spatial implications of changes in the regulation of South Africa's citrus export chain. In: Fold, N. and Nylandsted Larsen, M. (eds.) Globalization and restructuring of African commodity flows. Nordiska Afrikainstitutet, Uppsala, Sweden, pp. 79-102.

Mather, C. and Greenberg, S., 2003. Market liberalisation in post-apartheid South Africa: the restructuring of citrus exports after 'deregulation'. Journal of Southern African Studies 29: 393-412.

Memedovic, O., 2008. Editorial to special issue on global value chains and innovation networks: prospects for industrial upgrading in developing countries. International Journal of Technological Learning, Innovation and Development 1: 451-458.

Ndou, P., 2012. The competitiveness of the South African citrus industry in the face of the changing global health and environmental standards. PhD dissertation, University of Fort Hare, Alice, South Africa.

Neilson, J. and Pritchard, B., 2009. Value chain struggles: institutions and governance in the plantation districts of South India. Wiley-Blackwell, West Sussex, UK.

Obi, A. and Pote, P., 2012. Technical constraints to market access for crop and livestock farmers in Nkonkobe municipality, Eastern Cape province. In: Van Schalkwyk, H.D., Groenewald, J.A., Fraser, G., Obi, A. and Van Tilburg, A. (eds.) Unlocking markets to smallholders. Lessons from South Africa. Wageningen Academic Publishers, Wageningen, the Netherlands, pp. 91-112.

Perez-Aleman, P., 2011. Collective learning in global diffusion: spreading quality standards in a developing country cluster. Organization Science 22: 173-189.

Petzoldt, M., Joiko, C. and Menrad, K., 2008. Factors and their impacts for influencing food quality and safety in the value chains. University of Applied Sciences Weihenstephan, Freising, Germany.

Pietrobelli, C., 2008. Global value chains in the least developed countries of the world: threats and opportunities for local producers. International Journal of Technological Learning, Innovation and Development 1: 459-481.

Ponte, S. and Ewert, J., 2009. Which way is 'up' in upgrading? trajectories of change in the value chain for South African wine. World Development 37: 1637-1650.

Ponte, S. and Gibbon, P., 2005. Quality standards, conventions and the governance of global value chains. Economy and Society 34: 1-31.

Reardon, T., Timmer, C.P., Barrett, C.B. and Berdegué, J.A., 2003. The rise of supermarkets in Africa, Asia and Latin America. American Journal of Agricultural Economics 85: 1140-1146.

Ruben, R., Van Tilburg, A., Trienekens, J. and Van Boekel, M., 2007. Linking market integration, supply chain governance, quality and value added in tropical food chains In: Ruben, R., Van Boekel, M., Van Tilburg, A. and Trienekens, J. (eds.) Tropical food chains: governance regimes for quality management. Wageningen Academic Publishers, Wageningen, the Netherlands, pp. 13-46.

Saliola, F. and Zanfei, A., 2009. Multinational firms, global value chains and the organization of knowledge transfer. Research Policy 38: 369-381.

Selwyn, B. 2008. Institutions, upgrading and development: evidence from North East Brazilian export horticulture. Competition and Change 12: 377-396.

Symington, S., 2008. Creating sustainable competitive advantage in the marketing of South African table grapes to the United Kingdom in the deregulated era. MPhil dissertation, University of Cape Town, Cape Town, South Africa.

Van Dijk, F.E. and Maspero, E., 2004. An analysis of the South African fruit logistics infrastructure. ORiON 20: 55-72.

Vermeulen, H., Jordaan, D., Korsten, L. and Kirsten, J.F., 2006. Private standards, handling and hygiene in fruit export supply chains: a preliminary evaluation of the economic impact of parallel standards. In: International Association of Agricultural Economists Conference, 12-18 August 2006. Gold Coast, Australia.

Vermeulen, H., Kirsten, J.F. and Sartorius, K., 2008. Contracting arrangements in agribusiness procurement practices in South Africa. Agrekon 47: 198-221.

World Bank, 2007. World development report 2008. Agriculture for development. World Bank, Washington, DC, USA.

8. Towards achieving sustainable market access by South African smallholder deciduous fruit producers: the road ahead

B. Grwambi[1]*, P. Ingenbleek[2], A. Obi[3], R.A. Schipper[4] and H.C.M. van Trijp[2]

[1]Western Cape Department of Agriculture, Agricultural Economics Services, Marketing and Agribusiness Division, Private Bag X1, Elsenburg, 7607 South Africa; [2]Wageningen University, Marketing and Consumer Behaviour Group, Hollandseweg 1, 6706 KN Wageningen, the Netherlands; [3]University of Fort Hare, Faculty of Science and Agriculture, Department of Agricultural Economics and Extension, Private Bag X1314, Alice 5700, South Africa; [4]Wageningen University, Development Economics Group, Hollandseweg 1, 6706 KN Wageningen, the Netherlands; bukelwag@elsenburg.com

Abstract

Markets with high purchasing power, like export markets, offer opportunities for African smallholder farmers to move out of poverty. Logically, production for such higher value markets requires a different set of farm resources than the basic factors of production like land and labour. Yet it is not clear which other resources smallholder producers require to participate in different markets including high value markets. Using case studies from the Western Cape Province of South Africa, the authors identify resources that smallholder producers in developing countries require to increase competitiveness and sustain participation in high value markets. An analysis of the cases suggests that smallholder producers who are either in strategic partnerships or mentorship programmes with private sector firms are able to sustain their participation in high value markets. For smallholder producers who have not been integrated with high value markets yet, development of factor markets could be the first step towards achieving sustainable market access.

Keywords: resources, innovation, quality, strategic partnerships, high value markets

8.1 Introduction

Smallholder producers constitute the majority of the poor in developing countries and rely on agriculture for a living (IFC, 2011). Market access plays an essential role in the welfare of smallholder producers; by participating in markets, smallholder producers are able to earn an income which they use to meet household needs (Ehui *et al.*, 2003). As a result, initiatives aimed at reducing poverty in developing countries have been consistent with the pro-poor agricultural growth strategy and the market-led paradigm (Boughton *et al.*, 2007). Market access policies are becoming more important as new opportunities for increasing smallholder producers' incomes are increasingly being brought to the fore.

The changes in the global agricultural trade environment have created opportunities for the export of high value products by smallholder producers in developing countries (Delgado, 1999). Domestic and regional markets are now open and offer lucrative opportunities for smallholder producers to supply higher value produce and earn higher margins (Birthal *et al.*, 2007). When connected to high value markets, small and medium enterprises have more opportunities to upgrade and sustain income growth by moving into higher levels of value adding in the value chain (Humphrey and Schmitz, 2002). Connecting smallholder producers to high value markets thus serves as an indispensable pathway out of poverty and an important strategy to improve smallholder producers' livelihoods.

The extent to which smallholder farmers are integrated in markets has received attention from the literature on market integration. This literature approaches market integration of smallholder producers from the output side; it uses the quantity and/or value of output or product sold to measure the extent to which farmers are integrated in markets (Govereh *et al.*, 1999; Strasberg *et al.*, 1999). With food markets transforming from a market of commodities to a market of differentiated products heavily contested by powerful firms in consolidated food sectors, the literature is moving towards a perspective that appreciates farmers' ability to meet standards and preferences of customers. The institutional approach in which such institutions as contracts help farmers to add value to their products through compliance with standards has dominated the development literature on linking smallholder producers to high value markets.

With the opening of world markets to global trade where all producers, irrespective of their scale of operation and/or level of development compete for shelf space, adopting theories from developed countries regarding how firms deal with competition to maintain their market place positions in a changing market environment is a perspective worth exploring in studying smallholder market access challenges in developing countries. This chapter therefore builds on the Resource-Based View theory of the firm which distinguishes different categories of resources that contribute in different degrees to the long-run competitive advantage of companies (e.g. Eisenhardt and Martin, 2000; Hunt, 2000). It extends the literature on market integration by developing insights on how smallholder producers progress from generating low quality products to generating high quality products to meet customer demand in high value markets as they gain access to resources that have potential to increase their competitiveness and sustain their access to markets. We use qualitative research to develop a framework that conceptualises resources in smallholder producers' settings in developing countries. This framework is developed from the context of smallholder deciduous fruit producers in South Africa.

The South African context is insightful in that South African policy makers saw themselves confronted with a problem of persistent poverty and inequality skewed along racial lines while there were sufficient resources to develop policies that may help redress this imbalance. South Africa has now embarked on a process to review its policies and has, in its approach increasingly acknowledged a need for improving the access of smallholder farmers to land, water and institutional support systems as a means of combating poverty (Magingxa and Kamara, 2003).

The remainder of this chapter is structured as follows. A theoretical background which reviews literature on the participation of smallholder producers in high value markets and how the Resource Based View theory can contribute to addressing concerns of sustainability of market linkages between smallholder producers and buyers and of competitiveness of smallholder producers in high value markets follows. This is followed by an outline of the methods used to collect data. The chapter then describes the results which lead to the development of a model for integration of smallholder producers in high value markets that is potentially applicable to other contexts. The chapter draws conclusions which lead to implications for policy making in developing countries including South Africa. Directions for further research are also outlined at the end of this chapter.

8.2 Theoretical background

8.2.1 Participation of smallholder producers in high value markets

Further to the institutional approach followed in studying smallholder market access challenges in developing countries, the development literature has specifically studied participation of smallholder producers in high value markets. While positive impacts on incomes were reported (see for example Maertens and Swinnen, 2009; Minten et al., 2009), this literature raises two concerns which make the sustainability of participation of smallholder producers in high value markets doubtful.

The first concern relates to the competitiveness of smallholder producers in high value markets. Two strands of the development literature affirm to this. One strand emphasises the role of transaction costs in inhibiting small farmer participation in alternative marketing channels, including high-value markets (Jaffee, 1995; Key et al., 2000; Pingali et al., 2005; Staal et al., 1997), suggesting that the stricter (and increasing) requirements of high-value markets place smallholder producers at a competitive disadvantage relative to larger farmers. Another strand of the development literature emphasises that smallholder farmers' limited access to key resources (including cultivable land, irrigation and financial resources) inhibits investment (required for participation in high value markets) and farm productivity (McCulloch and Ota, 2002; Simmons, et al., 2005; Winter-Nelson and Temu, 2005). According to Henson et al. (2008), these assessments suggest that the lack of infrastructure, key production assets, information and/or collective action act to constrain initial smallholder entry to high-value markets and threaten the sustainability of such participation over time.

The second concern relates to the sustainability of market linkages between smallholder producers and buyers sourcing for high value markets. The view is that market leaders will only commit to sourcing from a particular base of smallholder producers while it is profitable to do so, yet will quickly 'jump ship' to new suppliers if problems arise and/or the opportunity arises to reduce their procurement costs (Goss *et al.*, 2000; Kolk and Van Tulder, 2006). Downstream buyers in high-value markets will choose to procure from sources that meet their requirements at lowest cost and at manageable levels of risk. They may prefer to source from larger-scale producers (where they exist) unless there are offsetting reasons to continue procuring from smallholder producers. Based on these possibilities, market arrangements made between smallholder producers and downstream buyers to access high value markets may not present a viable long-term approach to poverty alleviation (Henson *et al.*, 2008).

To address these concerns, this chapter applies a Resource Based View theory. The Resource Based View builds on theoretical insights from the debate on competitive advantage from strategic marketing and management literature. It maintains that possession of certain key resources developed over time is what assists firms to increase and maintain competitiveness in changing market environments. The Resource Based View theory thus offers insights on how resources developed over time (may) enable smallholder producers to increase their competitiveness and ability to participate in high value markets on a sustainable basis. An overview of the Resource Based View theory is presented below.

8.2.2 The resource based view theory of the firm

The Resource Based View explains how firms achieve and sustain a competitive advantage over time (Eisenhardt and Martin, 2000) and contends that sustained competitive advantage lies in the possession of certain key resources. Hunt (2000) defines resources as the tangible and intangible entities available to the firm that enable it to produce efficiently and/or effectively a market offering that has value for some market segments. For resources to be able to sustain competitive advantage: (1) they must be valuable in the sense that they exploit opportunities and/or neutralise threats in a firm's environment; (2) they must be rare among a firm's current and potential competition; (3) they must be imperfectly imitable; and (4) for the resources that are valuable but neither rare nor imperfectly imitable, there cannot be strategically equivalent substitutes (Barney, 1991).

According to Fahy (2000), a sustained competitive advantage can be obtained if the firm effectively deploys those resources in its product markets. Firms thus attempt to exploit valuable, rare, non-substitutable and inimitable resources in order to develop and sustain competitive advantages through their capabilities (Capron and Hulland, 1999). In rapidly changing environments, there is obviously value in the ability to sense the need to reconfigure the firm's asset structure and to accomplish the necessary internal and external transformation (Amit and Schoemaker, 1993; Langlois, 1994). The Resource Based View suggests firms can develop dynamic capabilities allowing them to build and reconfigure internal and external resources to address rapidly changing environments (Eisenhardt and Martin, 2000, Teece *et al*, 1997).

8.3 Methodology

8.3.1 Purpose

This study followed an exploratory case study research design. The purpose is to use theory to develop a conceptualisation of resources that smallholder producers in developing countries require to participate in different markets.

8.3.2 Research context

For reasons such as the strategic nature (e.g. export orientation and high labour requirements) of the subsector in terms of potential in realising the state's objectives of increased economic growth and job creation, we focused on the deciduous fruit subsector (NGP, 2011) in the Western Cape Province of South Africa. The Western Cape Province is the largest and traditional producer of deciduous fruit accounting for approximately 90% of the total production and exported apples (USDA, 2014). The industry creates approximately 110,000 job opportunities supporting 437,757 dependents (Hortgro Key Deciduous Fruit Statistics, 2014).

8.3.3 Case selection

We selected a farm as a unit of analysis because firms or farms in the case of farmers are considered in themselves as a bundle of resources (e.g. Penrose, 1959). Because firms build up and accumulate resources over time (e.g. Eisenhardt and Martin, 2000), we considered farms that have been established\acquired in earlier years as well as farms that have been established\acquired in recent years. As firms' resource base results to differentials in performances (e.g. Foss, 1998), we included various cases to capture differences in the resource base and the resulting performance differences.

We selected farms in which either unemployed adult individuals, regardless of whether they have farming experience or not, self-selected themselves to pool enough grants from the state to buy a farm with the hope to drive away poverty; or individuals working on a farm who were encouraged and supported by the commercial farmer they work for to apply for state grants to buy a separate farm. We refer to these farms as farms solely owned by smallholder producers. The study also included farms in which individuals who have been working on farms for most of their life time were given an opportunity to buy a share in the business of the commercial entity in which they work (i.e. joint ventures). We also studied farms in which competent individuals who have been longing to run farms of/on their own were given an opportunity to farm on state land on a rental basis (i.e. state farms) as well as farms acquired through other means (privately-owned farms).

These cases were selected from the historical and major fruit producing regions in the province. The study will show why some farms outperform others in terms of generating superior customer value hence ability to access high value markets.

8.3.4 Study questions

The general research question is why are some farms able to access high value markets while others are confined in low value markets? The more specific questions are:
1. To which markets do farms sell their produce?
2. Which resources enable farms to sell to these markets?
3. Which resources constrain farms from selling to other markets?

8.3.5 Data collection and analysis

Secondary data were collected from industry reports, commissioned studies, discussion papers, project reports and policy documents using desk research. We developed a draft case study protocol as recommended by Yin (2003) which was discussed with the team of investigators and experts who have a mandate to develop and transform the agricultural sector in South Africa. Primary data were collected through interviews with key informants. These respondents included researchers, public and private sector employees, buyers of fresh produce and technical and marketing specialists rendering advice to smallholder producers on a consultative basis. Buyers included an exporter, two fresh produce market agents and two retailers. The interviews with buyers took place in the packing firms, fresh produce markets and retail stores respectively which allowed the researchers some time to make observations. On a more informal and *ad hoc* basis, conversations were, in no particular order, held with the aforementioned informants. We also held a focus group discussion with five smallholder deciduous fruit producers. The discussion took approximately one and a half hours to conclude. The interviews with smallholder producers took place in the farms. This gave the researchers an opportunity to make site observations. Each interview lasted approximately one hour. The researchers recorded all interviews and then made full transcripts. The observations on the farms were made over time. The same cases that were studied were also visited by two investigators to study the same phenomenon. The focus group discussions and the visits by the different investigators enhanced external validity and added to reliability of the findings. Tables 8.1 and 8.2 offer an overview of respondents.

8.3.6 Case analysis

The team prepared four extensive case descriptions and, following Eisenhardt's (1989) suggestion, analysed within-case data independently as a stand-alone entity.

Table 8.1. List of buyers interviewed.

Buyer	Variety of produce sourced	Respondent's designation	Gender	Age	Education
1	deciduous fruit	exporter/accountant	male	33	degree in accounting
2	fruit, vegetables and other products	retail store manager 1	male	45	diploma in marketing
3	fruit, vegetables and other products	retail store manager 2	male	48	diploma in business management
4	fresh fruit and vegetables	market agent 1	male	30	master's degree in business administration
5	fresh fruit and vegetables	market agent 2	male	55	grade 12

Subsequently, three researchers conducted a cross-case comparison independently from each other, in order to find cross-case patterns. The research team then compared and discussed the patterns each individual researcher observed. The key components and relationships were discussed before systematically comparing these relationships with the original case data in order to re-examine the evidence from a new perspective, as an iterative process (Eisenhardt, 1989; Yin, 2003). The team distinguished two dimensions on which resources differ: (1) the level of tangibility of the resource which basically refers to its material nature, i.e. the physical existence of the resource which competitors can observe; and (2) the level of substitutability, i.e. how easy or difficult it is for competitors to obtain or acquire the said resource (Barney, 1991). From these two dimensions, we inductively developed a typology of four different bundles of resources as case descriptions presented below.

8.3.7 Typology of resources

'Factors of production' refer to such resources as land, labour, capital and water (Krugman, 2012). These resources are characterised by high levels of tangibility and can be easily bought in the market. This means, factors of production are highly substitutable and as such do not have the ability to sustain participation of smallholder producers in high value markets (Mathews, 2003).

'Personal intangible resources' on the other hand refer to more tacit and intangible resources such as skills and knowledge. For skills to be acquired, more practice is required. Employees who have acquired certain skills and knowledge can be hired (Nelson and Winter, 1982) which gives skills and knowledge moderate levels of tangibility and substitutability. As a result, personal intangible resources do not have the potential to sustain participation of smallholder producers in high value markets.

'Process capabilities' refer to key processes, such as production planning, quality control including execution of integrated pest management practices, logistics and marketing of produce that make use of factors of production and personal intangible resources of other chain actors to support farm activities. Due to complexity and interconnectedness of these processes and activities (Hunt, 2000), process capabilities have low levels of both tangibility and substitutability and thus have the potential

Table 8.2. List of farmers interviewed.

Farm	Year of establishment/ occupation	Farm size (ha)	Enterprises	Area under fruit (ha)	Labour[a]	Respondent's designation	Gender	Age	Education
1	2008	302	apples, pears, nectarines	144	34	production manager	male	57	grade 12 plus courses in fruit production
2	1999	200	apples, pears, wine grapes	36	14	farm manager	male	48	diploma in agriculture
3	2004	514	apples, pears, nectarines, peaches	274	932[b]	general manager	male	63	degree in agriculture
4	2005	669	apples, pears, peaches	106	34	farm manager	male	41	grade 12
5	2008	162	apples, pears	53	20	farm manager	male	70	diploma in agriculture
6	1999	26	peaches	4	2	farm manager	male	60	grade 10
7	2003	60	peaches, pears	12	4	farm manager	male	47	grade 12
8	2010	25	apples, lucer, livestock	7	3	farm manager	male	55	diploma in agriculture
9	2008	873	apples, pears, sheep	22	12	farm manager	male	50	diploma in agriculture
10	Early 1950s	40	apples, vegetables, livestock	4	3	farm manager	male	54	grade 12
11	Early 1950s	33	apples, vegetables, livestock	2	2	farm manager	male	55	grade 7
12	2010	1,049	apples, pears, peaches, plums, onions, potatoes, butternut, lucerne, sheep, cattle	8	12	farm manager	male	70	grade 8
13	Mid 1800s	3,200	apples, pears, peaches, onions, dry beans, oats, wild clover, sheep	27	15	farm manager	male	52	grade 12
14	1991	94	apples, pears, peaches, plums, apricots, nectarines, cabbage	37	40	farm manager	male	30	diploma in agriculture

[a] Figure covers only permanent workers.
[b] Number inclusive of pack house workers.

to enable smallholder producers to maintain their competitive positions in high value markets.

'Dynamic capabilities' refer to the ability of the farmer to make strategic changes in the resource base in order to keep up with the changing market environment (Teece, 2007; Teece *et al*, 1997). Examples include learning by doing, experimenting and risk taking (Eisenhardt and Martin, 2000; March, 1991; Kogut and Zander, 1992; Zollo and Winter, 2002). These are characterised by very low levels of tangibility and substitutability and thus have the potential to enable farmers to maintain their competitive positions in existing markets and to gain access to new markets in changing market environments.

8.4 Results

This section presents results from a case analysis of fourteen farms in the Western Cape Province of South Africa. The analysis also incorporates insights from the five buyers of deciduous fruit.

8.4.1 Markets for deciduous fruit in South Africa

Two forms of markets for deciduous fruit exist in South Africa; a formal and an informal market. The formal market mainly caters for overseas clients and is dominated by large export firms sourcing mainly from commercial farmers. The remaining fruit is sold locally to large retailers such as supermarkets. The informal market, on the other hand, serves the local market and is dominated by informal traders supplying a wide range of customers. The main suppliers of the informal market are smallholder farmers. The formal and informal markets also differ with regard to the level of value adding and quality requirements. Table 8.3 presents an overview of the formal and informal markets and their corresponding quality requirements.

Table 8.3. Markets used by smallholder producers and corresponding quality levels.

Markets	Value Adding	Quality
Informal markets (traders buy from the tree) for sale to various buyers	none as there are no grades, no standards; no assortments in terms of varieties	Low
Informal markets (traders collect) for sale to various buyers via a pack house	limited as there is some grading done but no standardisation; limited assortments in variety; traders do bulking and packaging for various buyers	Average
Local supermarkets	moderate and includes various assortments, size, consistency; reliability, some good agricultural practices based on growing programmes, packaging requirements	Good
Export markets (existing and new)	high and there are various assortments with specifications for size, consistency, reliability, good agricultural practices, good manufacturing practices, certification, labelling, traceability, ethical and environmental considerations, etc.	High

8.4.2 Contextualisation of resources and markets among smallholder deciduous fruit producers in South Africa

Smallholder farmers possess certain resources that enable them to sell to different markets. They also experience some resource constraints which limit them from selling to other markets. Below we discuss each resource typology and how it relates to quality of produce and consequently market access of smallholder deciduous fruit producers.

8.4.2.1 Factors of production, quality and market access

Local markets set basic minimum requirements with which their suppliers must comply prior to any exchange/transaction and, farmers whose fruit falls below minimum quality requirements have to seek alternative markets. Smallholder fruit producers in possession of factors of production, such as water, labour, spraying equipment and other machinery that are insufficient or inadequate, were, although skilled, only able to generate low quality fruit which was sold directly to informal traders (Cases 8, 11 and 12). Due to their tangibility, such factors of production as machinery and equipment are susceptible to ageing, theft and vandalism. Some farmers own machinery and equipment that is generally old and in some cases poorly maintained which exposes it to frequent breakdowns. In cases where such machinery breakdowns were experienced, fruit quality was affected and the fruit was eventually sold to informal traders.

Similarly, another farmer who has a link with a community pack house and originally sells fruit from his other farm via this pack house had his irrigation system in the state rental farm vandalised which consequently interfered with the restoration and flow of water. This farmer reported a record of poor yield and quality. 'In this seven hectare fruit farm, water was sufficient for only four hectares; this situation stressed the trees and resulted to few small apples being produced' (Case 8). The farmer viewed the fruit as not suitable for sale through the community pack house. As a result, the fruit was sold directly to informal traders. Lack of other factors of production and the influence it has on the quality of the produce has been experienced by one of the farmers and this is how he reflected: 'I delayed harvesting as I did not have a proper place to pack and store the fruit upon harvesting; the fruit dropped off from the trees on its own and resulted to low volumes and poor quality fruit' (Case 12).

The limited or insufficient factors of production among deciduous fruit producers not only resulted in low quality fruit but also affected the yields, thus resulting in low quantities of fruit being harvested. Interestingly, the data showed that farmers whose produce falls below minimum quality requirements could still sell their produce through initiatives by other chain members. In such cases, informal traders catering for a wide variety of consumers with diverse preferences seem to be the most suitable channel to which smallholder farmers could sell their fruit. For instance, informal traders would drive to the farms during harvesting season to buy fruit. They would bring along own transport, fruit picking and packing equipment and the labour to do the harvesting and packing of fruit. This initiative by informal traders enabled farmers to sell their fruit.

With regard to availability of factors of production, the state has reallocated these resources among smallholder producers as an attempt to redistribute wealth and reduce the poverty levels in the country. Through various land reform programmes, individuals and groups of smallholder producers were given grants to purchase land in which they are the sole owners, some used the grants to buy shares in operating farms and/or processing plants, while others lease farms from the government for production purposes. The aim of this programme was to redistribute land equally among society.

To facilitate rights to use of water resources, smallholder producers were granted user rights by being given a temporary water licence. These licenses entitle farmers to withdraw a given amount of water with a given quality from a river or canal. For farmers to be able to draw water from its source to the field, they need equipment. The state offers a subsidy scheme which finances a percentage of the off-farm water infrastructure development and distribution costs to smallholder farmers. The water pricing strategy also makes allowance for lower tariffs for smallholder farmers, with these increasing over a period of five years from issuing of a license or registration of water use to cover operational and management costs (Schreiner and Naidoo, 2000).

To make labour available to sectors, the labour market was also reregulated. This reform also provided for a sectoral determination of minimum wages for farm workers (Du Toit, 2003). Commercial farmers now compete with other farmers and with other rural and urban non-farm activities to attract and retain labour (Vaca, 2003). With regard to the provision of capital resources, a state-owned scheme aimed at providing micro and retail agricultural financial services on a large, accessible, cost effective and sustainable basis to communal land areas as well as smallholder producers was also introduced. This credit facility offers two products, viz. production loans as a means of bridging finance to cover production, processing, and marketing costs and equipment loans to finance the purchase of loose tools, small-scale plant and machinery, irrigation, and other farming equipment (NDA, 2007).

8.4.2.2 Personal intangible resources, quality and market access

Local community pack houses have basic minimum quality requirements which suppliers must fulfil prior to any exchange/transaction. Smallholder farmers whose fruit meets basic minimum requirements of community/private pack houses were able to sell their fruit to this channel. These farmers were endowed with both factors of production and personal intangible resources (Cases 6, 7 and 13). Yet, personal intangible resources may become obsolete. With the introduction of new and the continuous upgrading of existing technologies, production techniques improve and so should the technical skills and knowledge of farmers. For instance, during thinning, some farmers would leave more fruit on the trees than is required and sometimes too many fruits were left on young trees, which resulted in the bearing of unevenly-sized fruit. The importance of thinning in influencing fruit size and therefore quality was further stressed by the exporter: 'Fruit size can be controlled through hand thinning and is important in assessing the quality of fruit'.

With deployment of a combination of factors of production and personal intangible resources, smallholder farmers are only able to produce fruit of average quality which they sell to pack houses. However, farmers who received advice on new techniques for pruning and thinning observed an improvement in the quality of their fruit. These farmers were able to deliver more of the fruit to be sold as fresh as opposed to be sold for juicing and canning in the following season (Cases 6 and 7). The fruit is mostly sold to pack houses and arrangements are made to collect and deliver fruit to the pack house which then decides to which markets fruit will be sold depending on quality. While improvements in fruit quality have been witnessed by farmers who received technical advice, some believed that even if quality can improve, they still will not be able to supply export and domestic supermarkets as they cannot generate the quantities required in those markets due to the small size of the land (e.g. Case 8).

With regard to the development of personal intangible resources, the majority of farmers in the deciduous fruit industry were farm workers or have farming backgrounds and as such have acquired the necessary technical skills and knowledge required in farming. For those farmers who do not have such farming background, an acknowledgement of such a gap has been made and as a response, the government has introduced various mentorship programmes, which include a skills development component. The aim is to build capacity within the agricultural sector and to transfer technical and business skills to farmers (DAFF, 2009).

8.4.2.3 Process capabilities, quality and market access

Export markets and domestic supermarkets require producers to supply a differentiated quality level in which production techniques, spraying programmes, quantities to be produced and quality requirements as stipulated by buyers have been followed and fulfilled. Compliance with the different processes to be followed in order to produce and supply the right quantities and quality at the right time, as specified by the respective buyers, puts farmers in a better position to access high value markets (e.g. Cases 2, 4, 5, 10 and 14). Knowledge of preference for fruit varieties in the different markets, a personal intangible resource becomes crucial in this regard.

As the exporter alluded, 'Green, yellow, red and/or bicoloured apple varieties, for example, are preferred over other varieties in certain markets and knowledge of this diversity and preference in varieties is important for widening market scope'. This point received further emphasis from the exporter as well as from a market agent; 'Our suppliers need to always bear in mind that they are producing for the export and not the processing market' is how the exporter reflected, and the market agent commented, 'Urban and rural consumers have different preferences regarding the size and packaging of the fruit they buy'. As a result, farmers who are able to produce and supply the right quantities and quality as specified by the respective buyers gain access to these high value markets.

One farmer who has developed a relationship with a neighbouring commercial farmer explained, 'I have a seasonal verbal arrangement with a neighbouring commercial farmer who brings along his machinery and equipment, chemicals and labour to administer integrated pest management programme in our farm, during harvesting, the commercial farmer brings along his labour, fruit picking ladders and bins to pick-up first grade fruit which he packs in his cold store for sale to the export market' (Case 10). The exporter who is in a partnership with some of the farms surveyed cited their involvement pre-harvest as one of the factors that adds value to the quality of the fruit produced, 'To ensure good quality, we dispatch our technical team among our suppliers to observe activities and assist where necessary; and, as a result, we have grown in terms of productivity per unit area' the exporter commented. Visits by the technical team to ensure production of good quality fruit were also confirmed by another farmer who is also in a partnership: 'The technical advisors visit once a week to guide us in spraying and other related activities' (Case 9). The fruit is mainly exported with the remainder being sold to domestic supermarkets.

Commercial farmers who serve as mentors in state rental farms (e.g. Case 9) and in smallholder owned farms (e.g. Cases 4 and 5) as well as strategic partners in joint ventures (e.g. Cases 1 and 2) have been in the industry and on the export game for some time now and as a result have bought enough time to build and accumulate the necessary capabilities that are required to participate in export markets. Smallholder farmers who are in strategic partnerships and/or mentorship arrangements with commercial farmers and/or export firms are able to complement their capability base by tapping into these capabilities. Through these innovative arrangements, smallholder farmers are able to access the needed resources to ensure production of good quality fruit thus securing access to high value markets (Bitzer and Bijman, 2014).

While deployment of factors of production and personal intangible resources allows farmers to produce average quality, deployment of process capabilities in combination with factors of production and personal intangible resources enable farmers to upgrade quality to meet high value market standards. Strategic partnerships and/or mentorship arrangements have gained popularity and are increasingly being recognised as a model to facilitate inclusion of smallholder producers in high value markets. Government and the private sector (e.g. commodity organisations) across the agricultural sector have implemented various mentorship programmes over time. Ideally, mentors and/or strategic partners should transfer these skills and build these capabilities into participating smallholder farmers by involving them in farming activities and decision making processes. Being in these partnerships not only enables smallholder farmers to benefit from these process capabilities; they also get technical support and operational capital to buy inputs thus enabling them to access high value markets.

8.4.2.4 Dynamic capabilities, quality and market access

New markets, requiring a differentiated quality level that exceeds that of current buyers and has more potential for added value than what can be obtained from existing supply chains, could only be supplied if smallholder farmers have dynamic capabilities in addition to factors of production, personal intangible resources and process capabilities. In the production of deciduous fruit for new markets, smallholder farmers need to be flexible enough to respond effectively to changing consumer tastes and preferences. For example, to be able to keep up with these changes, smallholder farmers must be able to supply new varieties, adopt new food safety and quality standards, etc. as required by these new buyers. Meeting these requirements requires planting of new varieties, which must have been tested in advance (experimentation) before the actual large scale planting; adoption of new food safety and quality standards which requires some changes in the way things are done. Depending on the change needed, these alterations could be done at factors of production level, e.g. erection of structures for waste disposal; at personal intangible resource levels, e.g. training of labourers on the safe use of chemicals or at process capabilities levels, e.g. quality control through reduced chemical use during production (Figure 8.1). Some smallholder farmers, who were keen to make and learn from mistakes, willing to try

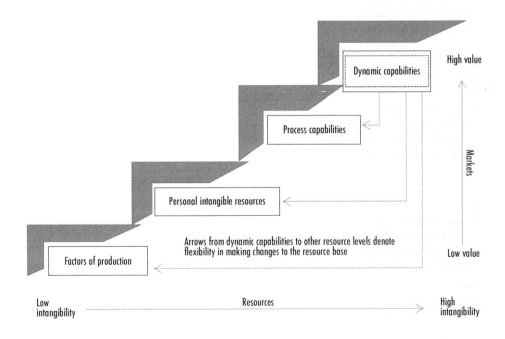

Figure 8.1. A model for integration of smallholder producers in high value markets.

out new things (risk taking), able to keep abreast of new developments and integrate these dynamic capabilities with factors of production, personal intangible resources and process capabilities, were able to remain in markets and to some extent, gain access to new markets (Cases 1, 3 and 9).

Some farmers believed learning is continuous and that there is, and will always be, room for improvement in whatever one is doing. As elaborated by one farmer, 'We had our yield heavily affected by sunburn in the previous season, we did things differently this season and have put a lot of attention on post-harvest activities such as pruning, orchard clearing, and fertilisation and have thus observed a lot of improvements in yield and quality' (Case 1). The changes in consumer preferences also require farmers to invest in new varieties. As one of the farmers who sits in fora where new varieties are discussed articulated his experience: 'I am testing new varieties in my farm and have been planting trial trees in the past few years which are growing well, this helps me see what works and what doesn't' (Case 9). According to this farmer, his access to markets has improved as a result of this larger variety in product. Some of these farmers attend trade fairs, export development workshops and other related programmes. As a result, they are well informed with what is happening in the industry and supply mainly export markets and domestic supermarkets. The importance of staying informed in the fruit industry was also remarked by the exporter: 'staying abreast with technology, such as rootstocks, add to improved fruit quality'.

Most of the farmers who shared these experiences are either in strategic partnerships or have mentorship arrangements with commercial farmers and/or export companies. Through these innovations with strategic partners, mentors or export companies, smallholder farmers are able to remain competitive in markets. This was confirmed by an exporter who also serves as a strategic partner to some of the farmers interviewed. According to the exporter, accessing markets is not a problem and sustaining competitive advantage is the goal for the farmers with whom they have forged partnerships: 'we have built our reputation in the industry through the production of good quality fruit; our only concern now is to seek growth opportunities through exploitation of new market outlets to increase our market scope'. Some smallholder farmers have also managed to get access to more market opportunities especially new markets in Africa, for example (Cases 3 and 9).

Further to the development of process capabilities through commercial farmers' long term involvement and participation in export markets, the withdrawal process of state support and the opening up of markets to global competition have enabled them (commercial farmers) to think strategically and find ways to survive in this dynamic market environment. Commercial farmers had to develop strategies to survive, one of which is the forging of strategic relationships with smallholder farmers, thus taking advantage of the opportunities presented through introduction of new policies in the agricultural sector. Smallholder farmers who are in strategic partnerships (e.g. Cases 1, 3) and/or mentorship arrangements with export firms (e.g. Case 9) are able to survive in this dynamic market environment by piggybacking on the dynamic capabilities of their strategic partners and/or mentors.

Building on the findings, the team developed a model for smallholder farmer integration in markets. A schematic overview of the resource levels at farm level which include factors of production, personal intangible resources, process capabilities and dynamic capabilities and the corresponding markets is presented in Figure 8.1.

8.4.3 Analysis of individual farms at the different levels of the resource hierarchy

Below we discuss other factors which may have, in some ways, further influenced the position that a farm holds in the hierarchy of resources.

8.4.3.1 Farms at the first level of the hierarchy

Cases 8, 11 and 12 could not move to higher levels of the resource hierarchy for various reasons. For Case 8, which is a state-owned farm, the lessee had potential to move to higher levels of the resource hierarchy, but could not do so due to water constraints. To keep the farm in running order, the farmer decided to incorporate a livestock enterprise. It is cases like this that prompted the state to introduce a programme that provides a fund to revive farms that are in a distressed state. As part of the norms and standards of this programme, the state makes it a prerequisite for such farms to have a strategic partner and/or a mentor before recapitalisation funds could be disbursed (DRDLR, 2011).

For Case 11, stepping up the hierarchy of resources might have not been a desirable option possibly due to difficult farming conditions prompted by numerous incidents in the previous decades (e.g. the oil crisis in the 1970s, the closure of the local railway station, politically motivated economic sanctions, etc.). These conditions led to the uprooting of trees and a subsequent diversification of farming activities by some smallholder producers in the community (Hart and Burgess, 2006) including Case 11. The farm currently runs a mixed farming system in which fruit is not the main enterprise.

Case 12, a privately owned farm could not move to higher levels of the resource hierarchy for two possible reasons. First, the farm's use of water that was not registered at the time of the interview may have influenced the farm's access to other tangible resources, such as credit. Second, the outright purchase instead of a lease from the state could mean that, the farmer's credentials in terms of managerial and farming experience were not convincing enough to warrant the state to lease him a farm.

8.4.3.2 Farms at the second level of the hierarchy

Cases 6, 7 and 13 were, for different reasons not able to move to higher levels of the resource hierarchy. For Cases 6 and 7, remaining on the farms and keeping the farms in production irrespective of harsh farming conditions, while other members left to seek alternative livelihood strategies, symbolises commitment and the desire to make a better livelihood out of farming. For reasons, such as non-compliance with food safety and quality standards, as well as fruit varieties that were only suitable for sale in the local market, Cases 6 and 7 could not move to the next level of the resource hierarchy. These farmers continue to farm under difficult conditions and one of the farmers reckons it is good that he kept his off-farm job as the income earned helps to at least keep the farm in running order.

Case 13 is stuck at the level of personal intangible resources, and infrastructural constraints may have kept this farm at this level. The remote location of the farm, the absence of electricity in the farm and the bad condition of the road to the farm which regularly causes damage to the fruit during transportation to the market may have limited growth of the farm. To take advantage of abundant water supplies, the farmer has included other enterprises, such as vegetables, but fruit remains the main enterprise in the farm. The farmer is eager to move to the next level of the hierarchy, such that they have started to make some improvements in terms of on-farm infrastructure in preparation for compliance with food safety and quality standards.

8.4.3.3 Farms at the third level of the hierarchy

Cases 2, 4, 5, 10 and 14 were not able to move to the top level of the hierarchy for reasons which vary from case to case. For Case 2, which is a joint venture, there are two possible reasons. First, the strategic partner's main focus is wine production and deciduous fruit was introduced as a new enterprise in the business. This could mean that the strategic partner had limited expertise in the production and marketing of deciduous fruit hence appointment of a farm manager and an exporting firm to attend to the production and marketing of deciduous fruit, respectively. Second, smallholder farmers are not shareholders in the company that packs and markets their fruit which may have implications regarding the extent to which the company invests in the farm.

With regard to Case 4, a smallholder owned farm which was initially a joint venture, various factors may have influenced the current position of the farm in the hierarchy of resources. The remote location of the farm and underdeveloped infrastructure in the area, the time and resources invested in the process of buying out the strategic partner from the business, the time between the termination of the relationship with the strategic partner and reinstatement of a mentor in the farm may have all in some way contributed to the delay of the farm in progressing to the top level of the hierarchy.

For Case 5, which is a smallholder owned farm, the old orchards including the time lag between their re-establishment and fruit bearing age of new trees may have contributed to keeping the farm in this position. Although in Case 10 the farmer has been getting support from the commercial farmer in the neighbourhood in the previous years, this farm remains at this level of the hierarchy. This is the case because it was never the objective of the commercial farmer to empower smallholder farmers in this historical deciduous fruit producing area rather than a process to add value to a product that will eventually increase his supply base. With recent developments in policies regarding empowerment of smallholder farmers, the state is also calling upon such commercial farmers to come forward to formalise these relationships. These are also the kinds of relationships in which suspicions of exploitation have been raised (DRDLR, 2011).

Case 14 is a private farm in which the owner is also involved in informal trading. The farm has moved to this level of the resource hierarchy after the farmer bought a share in the business of a firm which exports some of his produce. The farmer's informal trading background could be what has made him attractive as a potential business partner as the exporter could make use of the farmer' local networks to sell fruit not suitable for exports. It is unlikely that this farmer would abandon his traditional markets built over the years and focus exclusively on exports.

8.4.3.4 Farms at the fourth level of the hierarchy

For Cases 1 and 3, having strategic partners who run vertically integrated businesses, i.e. grower-packer-exporter structures, may have helped smallholder producers to stay abreast of latest developments in technologies and consumer trends. The strategic partners have also supported smallholder producers to respond to these changes in order to remain suppliers in high value markets.

For Case 9, which is a state-owned farm, a completely different observation has been made. Membership of the farmer in the company that markets his fruit entitled him to technical advice, rebates on packing and marketing costs, as well as financial advance for seasonal production; his extensive farming experience has given him an opportunity to serve in both technical and managerial positions as well as build relationships with various stakeholders in the industry; and his membership and service in various committees of farmer organisations, including his current position as one of the four farmers who are part of a reference group representing smallholder farmers at both provincial and national levels, might have given him enough exposure to think differently and act proactively. A combination of these experiences might have also contributed to enabling the farmer to step up and land himself to the top level of the hierarchy of resources.

Forging strategic partnerships and/or mentorship arrangements holds some benefits for all parties involved. For strategic partners and/or mentors, being in this kind of relationship with smallholder producers means an expanded supply base at minimal costs. For smallholder farmers, buying shares in a commercial entity was an opportunity worth taking as this meant capacitation in terms of exposure in a business environment and economic empowerment through participating in the mainstream economy. For the state, these partnerships meant development of a fledging class of smallholder producers capable of running viable and profitable farms in the near future which will eventually contribute to reduced levels of poverty in the country (DRDLR, 2011).

8.4.4 Case comparisons

The four resource typologies were compared and inferences as to how smallholder producers move from supplying low value markets to supplying high value markets that offer a higher margin than the former markets were drawn. The main distinguishing factor is the nature of the resources at the smallholder producers' disposal which relates to their potential to add value to the product and thus their ability to enable smallholder producers to maintain access in existing markets and to gain access in new markets.

The main difference between farms producing low quality and those producing average quality products is that, farms producing low quality products do not add any value to the product for reasons related to insufficient labour resources on the farm while those producing average quality products do add some value to the product. In those farms where no value is added, buyers come to the farm and use their own resources to add value to the product. For those smallholder producers who produce average quality products, the farmers at least harvest the crop and communicate with the buyer regarding collection of the crop upon harvesting. This way, some value addition is done at farm level; the pack house also adds value to the product through grading of produce. We therefore propose:

Proposition 1. Farms endowed with factors of production and personal intangible resources are only able to produce a low to average quality products that can be sold to low value markets.

Strategic partnerships and/or mentorship arrangements seem to be instrumental in the upgrading of quality and facilitation of linkages between smallholder producers and high value markets. Smallholder producers who have access to factors of production and personal intangible resources and who have forged strategic relationships with commercial farmers and/or export firms through either shareholding in joint ventures or mentorship arrangements in state and smallholder producer owned farms were made aware of what the market requires and received assistance during production to ensure compliance with the market requirements. Smallholder producers were also assisted with post-harvest handling activities such as packing and storage, and with marketing and distribution of the crop. We therefore propose:

Proposition 2. Farms endowed with factors of production, personal intangible resources and process capabilities are able to produce a good to high quality product that can be sold to high value markets, if strategic partners and/or mentors (1) share market information, and (2) render technical advice.

The ability to keep up with changes in consumer tastes and preferences, new developments in technologies and in the industry at large and the ability to quickly make alterations and amendments in response to those changes enables smallholder producers to continue participating in high value markets. Smallholder producers who are in strategic partnerships and/or mentorship arrangements with commercial farmers were able to get market information on current and future trends. The feedback from buyers also stimulated smallholder producers to be creative in terms of seeking and finding new ways to improve quality while long-term solutions are being sought.

Proposition 3. Farms endowed with factors of production, personal intangible resources and process capabilities are able to sustain their participation in high value markets if they have dynamic capabilities to keep abreast of market trends and latest technologies and bring this information to the level of smallholder producers to proactively act upon through (1) experimentation and learning and (2) adoption of new varieties.

8.5 Conclusions and policy implications

This study has identified resources which smallholder producers in developing countries require to participate in different markets. While fruit of a certain quality will find a place in the market, high value markets, which present producers with opportunities to earn higher margins for products in which value has been added, could only be accessed if producers have sufficiently developed process capabilities. Being able to participate in high value markets may further contribute to reduced levels of poverty. As not all farmers who were able to access high value markets had dynamic capabilities, these findings suggest that dynamic capabilities may, for now, not be necessary for all farmers to participate in high value markets. The study has also indicated that development of resources requires involvement and participation of various stakeholders in which both the public and private sectors have a role to play.

8.5.1 Implications for smallholder market integration policies

From this study, four implications for policy makers in developing countries emanate:

Firstly, as factors of production could be obtained in the open markets, development of regulatory systems to guide functioning of factor markets in developing countries may serve as a mechanism in the (re-)allocation of scarce resources. The development of factor markets will thus facilitate the allocation of such resources as land, labour, capital and water which tend to constrain smallholder producers from participating in markets.

Secondly, personal intangible resources such as skills and knowledge of individuals require more practice to develop. To facilitate the development of personal intangible resources, on-field training to enhance skills development as well as on-the-job training to build the knowledge base of smallholder producers, could be instrumental in the development of such resources.

Thirdly, process capabilities, such as production planning, quality control, logistics and marketing are a complex and interconnected process, the development of which requires processes to follow and involves a network of stakeholders to liaise and relate with. Development of process capabilities by smallholder producers could be facilitated through involvement in all activities and strategic decision making processes of the business starting from production planning, quality control and logistics until the product reaches the market.

Fourthly, the development of dynamic capabilities is a long term process that builds on farming and business experience accumulated over time and plays a crucial role in sustaining participation of smallholder producers in high value markets. Support to the development of dynamic capabilities may, in combination with consistent market research cover continuous awareness creation in trade regulations, exposure trips to explore new technologies and new ways of doing things, timely updates on (new) standards development, and creation of opportunities and platforms for networking, experimenting, experiential learning and experience sharing. An institutional environment that provides insights in market development will also be instrumental in the development of dynamic capabilities.

8.5.2 Implications for South African policies

South Africa has made progress with regard to developing policies aimed at improving participation of smallholder producers in high value markets. The country seems to be also steering its policies towards the right direction as it has, in its policy development approach, already adopted the logic of a resource hierarchy.

South Africa has implemented various policy reforms to facilitate the functioning of its factor markets. The reforms were implemented in land, labour, capital and water markets. The land reform programme, labour market reform, development of micro-finance institutions and water reform programmes were implemented to redistribute and make these factors of production available to smallholder producers who were previously denied access to these resources.

For building and broadening the skills and knowledge base of smallholder producers, various strategic partnership and mentorship programmes specifically designed to bridge the skills and knowledge gap of smallholder farmers have been launched and implemented. Further to skills and knowledge development, these partnership and mentorship programmes are envisaged to capacitate farmers in terms of process capabilities over time. This is where South Africa currently stands in terms of policy development.

A policy recommendation for South Africa would be to strengthen and extend its monitoring systems at farm level to include an audit of skills and capabilities and where gaps are identified, develop action plans to fill the gaps. This will serve as a means to fast-track the development of process capabilities by smallholder producers.

More precisely, monitoring and evaluation systems at farm level could focus more on human resources and assess their level of skills development on a continuous basis as the mentorship programme progresses than focusing on tangible measures/indicators such as increases in incomes. This is not to say the latter measures/indicators are not important. This process is crucial as development of process capabilities will enable smallholder producers to move to the next level of resource development, which will sustain their participation in high value markets.

8.6 Directions for further research

This study underscored that market access of smallholder deciduous fruit producers is explained by the type and level of resources in possession of farmers. Future research needs to develop measurement instruments to test the conceptual framework and hypotheses which were developed in this study. Future research may examine the influence of resources on the market access of smallholder deciduous fruit producers, by exploring possible moderating variables that strengthen or weaken the relationship between resources and market access of smallholder farmers. Finally, longitudinal studies could provide insight on whether dynamic capabilities lead to sustained competitive advantage in the long-term in this context.

References

Amit, R. and Schoemaker, P.J., 1993. Strategic assets and organisational rent. Strategic Management Journal 14: 33-46.

Barney, J., 1991. Firm resources and sustained competitive advantage. Journal of Management 17: 99-120.

Birthal, P.S., Jha, A.K. and Singh, H., 2007. Linking farmers to markets for high value agricultural commodities. Agricultural Economics Research Review 20: 425-439.

Bitzer, V. and Bijman, J., 2014. Old oranges in new boxes? Strategic partnerships between emerging farmers and agribusinesses in South Africa. Journal of Southern African Studies 40: 167-183.

Boughton, D., Mather, D., Barrett, C.B., Benfica, R., Abdula, D., Tschirley, D. and Cungura, B., 2007. Market participation by rural households in a low-income country: an asset-based approach applied to Mozambique. Faith and Economics 50: 64-101.

Capron, L. and Hulland, J., 1999. Redeployment of brands, sales forces, and marketing expertise following horizontal acquisitions: a resource-based view. Journal of Marketing 63: 41-54.

Department of Agriculture, Forestry and Fisheries (DAFF), 2009. 2008/09 Annual report on the implementation of the commodity-based master mentorship programme. Department of Agriculture, Forestry and Fisheries. Pretoria, South Africa.

Delgado, C., 1999. Sources of growth in smallholder agriculture in Sub-Saharan Africa: the role of vertical integration of smallholders with processors and marketers of high value items. Agrekon 38: 165-189.

Department of Rural Development and Land Reform (DRDLR), 2011. Policy framework for the recapitalisation and development programme of the department of rural development and land reform. Department of Rural Development and Land Reform. Pretoria, South Africa.

Du Toit, A., 2003. Hunger in the valley of fruitfulness, globalisation, 'social exclusion' and chronic poverty in Ceres, South Africa. In: Conference 'Staying poor: chronic poverty and development policy', University of Manchester, Manchester, UK. April 7-9, 2003.

Ehui, S., Benin, S. and Paulos, Z., 2003. Policy options for improving market participation and sales of smallholder livestock producers: a case of Ethiopia. International Conference on African Development Archives, paper 77. Available at: http://tinyurl.com/jt7lxre.

Eisenhardt, K.M., 1989. Building theories from case study research. Academy of Management Review 14: 532-550.

Eisenhardt, K. and Martin, J., 2000. Dynamic capabilities: what are they? Strategic Management Journal 21: 1105-1121.

Fahy, J., 2000. The resource based view of the firm: some stumbling blocks on the road to understanding sustainable competitive advantage. Journal of European Industrial Training 24: 94-104.

Foss, N., 1998. The resource-based perspective: an assessment and diagnosis of problems. Scandinavian Journal of Management 14: 133-149.

Goss, J., Burch, D. and Rickson, R., 2000. Agrifood restructuring and third world transnationals: Thailand, the C.P. group and the global shrimp industry. World Development 28: 513-530.

Govereh, J., Jayne, T.S. and Nyoro, J., 1999. Smallholder commercialization, interlinked markets and food crop productivity: cross-country evidence in eastern and southern Africa. Available at: http://tinyurl.com/p8srvx2.

Hart, T. and Burgess, R., 2006. Across the divide: the impact of farmer to farmer linkages in the absence of extension services. In: Dixon, J., Neely, C., Lightfoot, C., Avila, M., Baker, D., Holding, C. and King, C. (eds.) Farming systems and poverty: making a difference. Proceedings of the International Symposium of the 18th International Symposium of the International Farming Systems Association (IFSA): a Global Learning Opportunity. IFSA, Rome, Italy. Available at: http://tinyurl.com/z84xn8o.

Henson, S., Jaffee, S., Cranfield, J., Blandon, J. and Siegel, P., 2008. Linking African smallholders to high value markets: practitioner perspectives on benefits, constraints and interventions. Policy Research Working Paper 4573. World Bank, Washington, DC, USA.

HortGro, 2014. Key deciduous fruit statistics 2014. HortGro Services, Paarl, South Africa. Available at: http://tinyurl.com/o7vaqrs.

Humphrey, J. and Schmitz, H., 2002. How does insertion in global value chains affect upgrading in industrial clusters? Regional Studies 36: 1017-1027.

Hunt, S.D., 2000. A general theory of competition: resources, competences, productivity, economic growth. Sage Publications, Thousand Oaks, CA, USA.

International Finance Corporation (IFC), 2011. Scaling up access to finance for agricultural SMEs: policy review and recommendations. International Finance Corporation, Washington, DC, USA. Available at: http://tinyurl.com/zjpl6uu.

Jaffee, S., 1995. Marketing Africa's high value foods-comparative experiences of an emergent private sector. Kendall/ Hunt, Washington, DC, USA.

Key, N., Sadoulet, E. and De Janvry, A., 2000. Transaction costs agricultural household supply response. American Journal of Agricultural Economics 82: 245-259.

Kogut, B. and Zander, U., 1992. Knowledge of the firm, combinative capabilities and the replication of technology. Organization Science 3: 383-397.

Kolk, A. and Van Tulder, R., 2006. Poverty alleviation as business strategy? Evaluating commitments of frontrunner multinational corporations. World Development 34: 789-801.

Krugman, O.M., 2012. International economics: theory and policy. The Heckcher Ohlin model. Pearson-Addison-Wesley, Boston, MA, USA.

Langlois, R.N., 1994. The market process: an evolutionary view. In: Boettke, P.J. and Prychitko, D.L. (eds.). The market process. Edward Elgar, Aldershot, UK, pp. 29-37.

Magingxa, L.L. and Kamara, A.B., 2003. Institutional perspectives of enhancing smallholder market access in South Africa. In: 41st Annual Conference of the Agricultural Economics Association of South Africa (AEASA), October 2-3, 2003, Pretoria, South Africa.

Maertens, M. and Swinnen, J., 2009. Trade, standards and poverty: evidence from Senegal. World Development 37: 161-178.

March, J.G., 1991. Exploration and exploitation in organizational learning. Organisation Science 2: 71-87.

Mathews, J.A., 2003. Strategizing by firms in the presence of markets for resources. Industrial and Corporate Change 12: 1157-1193.

McCulloch, N. and Ota, M., 2002. Export horticulture and poverty in Kenya. Working Paper 174. Institute of Development Studies, Brighton, UK.

Minten, B., Randrianarison, L. and Swinnen, J.F.M., 2009. Global retail chains and poor farmers: evidence from Madagascar. World Development 37: 1728-1741.

National Department of Agriculture (NDA), 2007. MAFISA credit policy framework. Policy Document. National Department of Agriculture. Pretoria, South Africa.

Nelson, R.R. and Winter, S.G., 1982. An evolutionary theory of economic change. The Belknap Press of Harvard University Press, Cambridge, MA, USA.

New Growth Path (NGP), 2011. New growth path policy document. Department of Economic Development. South Africa. Available at: http://tinyurl.com/pl9xujp.

Penrose, E.T., 1959. The theory of the growth of the firm. John Wiley, New York, NY, USA.

Pingali, P., Khwaja, Y. and Meijer, M., 2005. Commercialising small farms: reducing transaction costs. In: The future of small farms. Proceedings of a Research Workshop. International Food Policy Research Institute. Washington, DC, USA. Available at: http://tinyurl.com/hqo56mh.

Schreiner, B. and Naidoo, D., 2000. Water as an instrument for social development in South Africa. Paper presented at the 2000 Water Institute of Southern Africa (WISA) Biennial Conference. May 28–June 01, 2000. Sun City, South Africa. Available at: http://www.ewisa.co.za/literature/files/269schreiner.pdf.

Simmons, P., Winters, P. and Patrick, I., 2005. An analysis of contract farming in East Java, Bali, and Lombok, Indonesia. Agricultural Economics 33: 513-525.

Staal, S., Delgado, C. and Nicholson, C., 1997. Smallholder dairying under transaction costs in east Africa. World Development 25: 779-794.

Strasberg, P.I., Jayne, T.S., Yamano, T., Nyoro, J., Karanja, D. and Strauss, J., 1999. Effects of agricultural commercialisation on food crop input use and productivity in Kenya. Michigan State University International Development Working Papers No. 71. Michigan, USA.

Teece, D., Pisano, G. and Shuen, A., 1997. Dynamic capabilities and strategic management. Strategic Management Journal 18: 509-533.

Teece, D., 2007. Explicating dynamic capabilities: the nature and micro-foundation of (sustainable) enterprise performance. Strategic Management Journal 28: 1390-1350.

United States Department of Agriculture (USDA), 2014. South African deciduous fruit production forecasted to increase in the 2014/15 Mid-Year. Fresh Deciduous Fruit Annual. Global Agricultural Information Network (GAIN) Report. United States Department of Agriculture Foreign Agricultural Service. Available at: http://tinyurl.com/q4lykf2.

Vaca, J.F.A., 2003. Equity schemes in South Africa: benefits delivered to farm workers as function of commercial farmers' strategy. A Plan B Paper. MSc thesis, University of Michigan, Ann Arbor, MI, USA.

Winter-Nelson, A. and Temu, A., 2005. Liquidity constraints, access to credit and pro-poor growth in rural Tanzania. Journal of International Development 17: 867-882.

Yin, R.K. 2003. Case study research: design and methods. Sage Publications, Thousand Oaks, CA, USA.

Zollo, M. and Winter, S.G., 2002. Deliberate learning and the evolution of dynamic capabilities. Organisation Science 13: 339-351.

9. Institutional co-innovation in value chain development: a comparative study of agro-export products in Uganda and Peru

A.H.J. Helmsing* and W. Enzama

Erasmus University Rotterdam, International Institute of Social Studies, P.O. Box 29776, 2502 LT, The Hague, the Netherlands; helmsing@iss.nl

Abstract

Institutional development has attracted more attention in the past two decades. However, institutional theory finds itself in a pre-consolidated phase and there are many theoretical and methodological challenges. One is to respond to the question whether institutional change is a spontaneous evolutionary or a deliberately designed process or a combination of the two. Another question concerns the interaction between technological innovations, changes in institutional arrangements and changes in the institutional environment in the dynamics of processes of institutional development. This links to another key question concerning the synchronicity in or co-evolution of institutional change processes at various levels and in various public and private domains. Institutional innovation rarely concerns one single institution but normally concerns bundles of public and private order institutions created at various levels. This paper researches how a common institutional need to develop institutional arrangements for rural collective action in order to enable small farmers to participate in newly created export chains, each with its own technological requirements and in different contexts leads to different institutional arrangements and outcomes. By comparing two cases, the paper seeks to unravel which factors and actors play what roles and how these explain differences in the process of institutional development and in that way to arrive at a better understanding of local institutional change. After a review of literature and the elaboration of a framework to answer the above questions, the paper presents a bird's eye views of the two case studies. The first refers to the introduction of new apicultural technologies in the North West of Uganda and the second relates to the introduction of high value horticulture exports crops in the North of Peru. The final section examines the main commonalities and differences in institutional development and makes an attempt to respond to the main questions formulated above.

Keywords: institutional development, co-innovation, collective action, value chains, developing countries

9.1 Introduction

Since the late eighties, institutions have been recognised as playing an important role in economic development. It began with a critique of structural adjustment policies which centred too much on 'getting the prices' right, rather than on 'getting the institutions right'. Since the 1990s theorising about institutions has taken considerable leaps forward but still finds itself in a pre-consolidated stage and there are many theoretical and methodological challenges. Most attention has been given to influence of particular institutions on economic development and not on the reverse causality. Chang (2010) has made this point most effectively. Economic development also changes institutions as it gives rise to new agents and activities that demand new kinds of institutions; the wealth created in the process demands institutional change towards more accountability and transparency but also makes institutional change affordable.

A methodological challenge of comparative research in this regard is not to focus on an a-priori defined specific nominal institution but to focus on a common institutional need, which may give rise to distinct institutional solutions in distinct contextual settings (Maseland, 2011). In that way the interaction between institutions and economic development can better be captured. In this context, a key question is whether institutional change is one of deliberate design or a spontaneous evolutionary process (Kingston and Caballero, 2009) or a combination of these. The latter links to another key question, not raised by these authors, namely, concerning the synchronicity in or co-evolution of institutional change processes at various levels and in public and private domains. Institutional innovation rarely concerns one single institution but normally concerns bundles of public and private order institutions created at various levels. The literature often gives considerable attention to the State which is to provide an appropriate business institutional environment (i.e. set of institutions) within which economic agents and activity can prosper and within which private agents can develop their own complementary private order institutions. Is this necessarily a downward process where public institutions provide the framework within which private order institutions are created or adapted? Can the reverse also happen and if so, under what conditions? How do national and local level institutional change agents interact? Below we will give some conceptual elaborations necessary to answer these questions empirically.

We will then, in Section 9.2, give a bird's eye view of the two cases. The first concerns the introduction of new beekeeping technologies in the West Nile region of Uganda where a private company played a key role in creating an agro-export chain and non-governmental organisations (NGOs) a complementary one and the other case concerns the introduction of an agro-export crop in the Department of La Libertad in Northern Peru by a NGO with a private company as ally. The two cases constitute very different cultural and historical institutional settings (the state being more prominent in Uganda, than in Peru) but cover roughly the same period (2001-2008). In both instances the Government attempted to introduce market based agricultural business development services (BDS). The purpose of the analysis is not to identify a 'superior' institution to be replicated elsewhere but to get better insight in what complex set of factors and actors shape institutional change around a common institutional need.

The analysis of the two cases is structured using a time line. Necessarily, the presentation of the two cases will be sketchy and cannot be elaborated in all their richness for reasons of space. Section 9.3 concludes the paper and examines commonalities and differences between the two cases and contains some final observations concerning synchronicity in co-innovation processes.

9.1.1 Understanding institutional change and co-evolution of institutional innovations

Institutions are defined in a variety of ways in the literature. A common definition, states that institutions are rules of the game (see also Chang (2002). North (1990) maintained that institutions are humanly devised constraints that shape human interaction and Hodgson (1988, 2006) sees them as durable patterns of human interaction. Here we use the notion of rules, not as constraints but as rules that guide human behaviour. Nelson sees institutions as social technologies or 'ways of getting things done when human interaction is needed' (Nelson, 2008; Nelson and Sampat, 2001). Social technologies become institutions when they have become standard and the expected thing to do, given objective and setting (*ibid.*, 2001: 40). The literature identifies different kinds of institutions. The most common distinction is between formal and informal institutions where the former are often associated with written rules. A partially overlapping distinction states that specialised actors or organisations (including judges and courts) enforce formal institutions (like laws). Informal ones are endogenously enforced by members of the associated group (Kingston and Caballero, 2009). Institutions may be voluntary and constitute a private order, while public institutions are normally apply to all citizens or functional groups within a designated jurisdiction or functional area.

Institutions tend to be nested and hierarchical. 'Nestedness' refers the fact that institutions are interrelated and that institutions at one level, set the stage for institutions at another level. For example, Williamson (2000) identifies four types of institutions where the time horizon of change is taken as a key criterion: (1) institutions of embeddedness, including informal institutions and norms, change in the order of centuries or millennia; (2) high level formal rules such as constitutions, laws and property rights normally change in the order of decades; (3) institutions of governance set the rules for day-to-day interactions and can be modified in the short run; finally; (4) transaction contracts, which set prices and quantities, change continuously. New institutional economists do not study type (1) institutions but take them for granted. They focus on level (2) and (3) institutions.

Embeddedness of institutions operates in two ways: it implies that for operational rules, higher level rules can be taken as given or as exogenous; but also that for changing operational rules it may be necessary to change also higher level rules. This is one of the causes of institutional path dependence or inertia (see below). The 'nestedness' also explains hierarchy between different kinds of institutions, but there is also hierarchy between rules of the same kind. The constitution takes precedence over any other law but also some laws take precedence over other laws (e.g. criminal over commercial law or children rights over employer rights). The hierarchy of rules is one of the important issues of struggle between groups who stand to gain or lose from particular rules (Chang, 2002).

One aspect of hierarchy concerns the scale at which an institution applies. Does it refer to all economic agents or to (self-)selected groups? Do agents have alternative options to deal with a similar institutional need? In some theoretical approaches the process by which the scale of institution rises, is called 'climbing the institutional ladder' (see below).

What drives institutions to change? North (2005) places emphasis on the advances of cognitive science or the ways in which we interpret the world and its problems. Hodgson (2006) and Gomez (2008) assume that an institutional gap emerges as a result of endogenous or exogenous shocks. That is to say, 'when action X does not result in expected outcome Y'. Other authors like Nelson (2002; Nelson and Sampat, 2001) see technology as the principal driver of institutional change. A new physical technology (i.e. new ways of producing, new products or new ways of undertaking activities) calls for a change in (internal) routines within the firm and may call for a change in ways of doing things between economic agents (institutions). Nelson thus stresses the importance of co-innovation (in technology, organisation, marketing, etc.) and by implication institutional innovations in various domains.

If we follow this reasoning then institutional change involves various actors. The literature puts emphasis on two categories of actors, from two different domains: one is the catalytic or institutional entrepreneur who acts on his/her own account or in association with other entrepreneurs who face similar institutional needs. They may form an association and pursue their institutional interests collectively. The second actor is the state. Some authors stress the 'embedded autonomy' of the state to design new institutions through deliberate economic policy (Evans, 1995). Conceptions of state vary, as does the emphasis on specific state actors and aspects of state functioning and performance. Some authors distinguish between roles of state bureaucrats and/or politicians in the institutional design process. With regard to late developing countries, we could add multilateral and bilateral aid agencies, which influence the direction of institutional change. Others stress the interaction between the public and the private domains. The State not only plays a role in terms of designing and enforcing public institutions but it also provides legitimacy to private order institutions. Economic interest groups struggle for power to control the state and in that way influence the direction of institutional change (Chang, 2002).

New Institutional Economics (NIE) has made important contributions to the economic dimension of institutional change (Williamson, 1975, 2000). It is argued that under conditions of competitive markets durable institutional change occurs only when the new institution is efficient. Institutional options that result in fewer transaction cost reductions will eventually give way as being inefficient. At the micro level, NIE also points to the importance of asset specificities, information asymmetries, adverse selection and moral hazard problems, which trigger specific institutional designs. In many late developing countries markets are far from perfect; thin markets and market failures may be endemic and therefore such problems are far more frequent and as a result inefficient institutions may continue to abound.

As regards the sustainability of institutions, two perspectives tend to predominate. One emphasises the existence of competing institutional options and the voluntary nature of acceptance and replication. The instigator of a new rule may have to compensate others who may stand to lose from the new rule in order to prevent their opposition or their switching to another institutional alternative. The other stresses the role of the state and public enforcement of compliance. In this case the cost of enforcement is often not seen as an overriding consideration, as it remains hidden in overall cost of government but it does imply that administrative and public policy considerations may have considerable bearing on institutional design process. Bureaucrats may have a different view of the world than do entrepreneurs or politicians.

Institutional change is a path dependent process. Two important factors are situated bounded rationality of the actors and institutional inertia. Situated bounded rationality occurs, as learning and search processes are localised, time consuming and costly; as a result, information tends to be incomplete and processing capacity is limited. Actors therefore reveal satisficing behaviour. They will accept a 'good enough' institutional option. Risk aversion in situations of low trust and failing markets can further compound the selection process and prevent institutional change from becoming sustainable. North (2005) stressed the importance of mental models and ideologies with which actors work and which can influence the perceptions about the effects of alternative institutions. Institutional inertia can result due to the existence of free rider problems, which prevent collective action to change institutions. Furthermore, informal institutions can be an important source of institutional inertia. Last but not least institutional complementarities or interrelatedness can act as a severe brake on institutional change. Not only in functional terms but also because certain groups who benefit from the complementary institution can oppose the institutional change desired by others. In all, history or context plays an important role in the process of institutional change.

Brousseau and Raynaud (2007) have made an interesting contribution to the analysis of institutional change processes. They argue that institutional change begins as decentralised process of creating localised institutional orders. Local institutions tend to be voluntary as agents have always the option to create or join alternative institutional solutions to their common coordination need. Local institutional orders are therefore in competition with each other. They argue that at the local level economic competition takes place between different and alternative institutional options but at higher and more centralised level political competition between higher order institutions takes precedence. Their central contention is that local institutional arrangements have a built in tendency to seek to expand and 'climb the institutional ladder' and 'like lava, have the tendency to spread out and then solidify to become part of the institutional framework'. When competing, agents can deploy different strategies. They can adapt and improve the quality of the institution so as to improve its efficiency and relevance to others; poach of go-betweens (that is, agents that ascribe to a rival institutional order but with coordination needs that are close to one's own); manipulate switching costs, by making it less attractive for existing members to defect; retaliate ex-post and ostracise defectors; and lastly, negotiate a merger (Brousseau and Raynaud, *ibid.*). Timing and 'first mover' advantages thus play an important role in the process of institutional change.

The nature of the relationships between public and private actors constitutes a key aspect in the synchronisation of institutional change processes. Do higher public order institutions (have to) precede the creation of private order institutions? Alternatively, do private order institutions 'climb the institutional ladder' as Brousseau and Raynaud suggest and/or demand or give rise to required complementary new public institutions, as Chang argues? Or do these two public and private institutional innovation processes take place simultaneously? Is the synchrony a matter of deliberate coordination between levels and domains or is it a matter of chance as their respective path dependent processes co-evolve?

We began with two main questions: Can institutional change best be seen as spontaneous rather than an outcome of deliberate design or as a combination of these two; and, if institutional change is about a complex set of institutions co-evolving at different levels and in different domains, what influences the co-innovation of institutions? We have elaborated above concepts that can help us answer these questions empirically.

9.2 Bird's eye views of two case studies

9.2.1 Introduction

The two case studies have been monitored over a period of time and the analysis is based on multiple site visits. In both cases both chain actors (including small farmer groups) and chain promoters have been interviewed.

9.2.2 Institutional arrangements for beekeeping in West Nile, Uganda

In the 1970s and early 1980s Uganda suffered from considerable political turmoil and economic mismanagement but with the National Resistance Movement (NRM) of Museveni taking control, the economy rebounded and political stability improved. The average annual rate of growth in the nineties was 6%. The population below the poverty line declined from 56% in 1992 to 38% in 2003. But in Northern Uganda unrest was not contained and the Lord's Resistance Army (LRA) and the Uganda National Rescue Front (UNRF II) in West Nile continued to create havoc. Only in 2002 the Government signed a peace treaty with UNRF II rebel groups and post-war reconstruction began in West Nile.

In 2000 the Government of Uganda launched an ambitious new Policy for the Modernization of Agriculture (PMA) to enhance production, competitiveness and incomes. One of the seven pillars of the reform was the delivery of agricultural extension through a new National Agricultural Advisory Services (NAADS). NAADS is considered to be an innovative public-private extension service delivery approach with the aim to increase commercial farming among Uganda's subsistence smallholders. This program was officially launched in 2001 to promote the development of farmer organisations and empower them to (1) procure advisory services; (2) to manage linkages with marketing partners and (3) to conduct demand driven monitoring and evaluation of advisory services and their impacts.

Under the NAADS policy, farmers can form groups, negotiate with private sector (NGOs) service providers, and award short-term contracts to promote specific farm enterprises and provide advisory services (Benin, 2007). NAADS implements and manages its program at the sub-county (LC3) level. At this level, priority farm products are identified and NAADS manages from here the allocation of contracts, monitors and evaluates performance and accountability of service providers and farmer groups. At LC3 level farmer forums are established, composed of representatives of farmer groups, which themselves operate at village level (LC1). The farmer groups are the basic unit receiving the advisory services. Members are selected from among the economically active poor (i.e. neither the destitute poor nor larger scale farmers). They are encouraged to work together around a particular crop or farm enterprise. The farmer groups are given advice on how to organise themselves and engage in collective action (e.g. learning how to set themselves up as a farmer group with a constitution and how to make bye-laws, etc.), engage the local government and service providers, manage technical development sites and organise demonstration and training sessions.

Arua District in West Nile (Northern Uganda) was one of the six 'trail blazing' districts in which the NAADS program was initiated in 2001. It was rolled out in 24 sub-counties of these six districts. In 2002/2003 the program was extended to 10 additional districts and in 2003/2004 and 2004/2005 another 13 districts were incorporated. In 2005 NAADS was active in 29 (of the then total of 70) districts and 280 sub-counties with some 13,200 operating farmer groups (Benin, 2007). In West Nile beekeeping was selected as a promising farm enterprise alongside with the introduction of a new breed of goats and a new groundnut variety. Table 9.1 provides a timeline of the events in West-Nile.

Table 9.1. Timeline of events in West Nile.[1]

Year	Event[2]
1997	Local governments assume responsibility for agricultural extension services; staff transferred from national level to the districts
2000	Policy for the modernisation of Agriculture
2001	National Agency for Agricultural Extension
2001	Initiation of NAADS in six 'trail blazing' districts, including Arua
2002	Bee Natural Products (BNP) Ltd. founded in Kampala
2002	Comb honey processing plant opened by BNP Ltd. in Arua
2002	Peace accord signed for Northern Uganda
2003/5	NAADS expands operations in 13 additional districts
2003	UNIDO begin beekeeping project for ex-combatants
2003	Agreement between NAADS and BNP Ltd.
2005	Ugandan honey certified for EU market
2007	Second agreement between NAADS and BNP Ltd.
2008	Factory in Arua closed
2008	New processor starts in Yumbe
2009	Factory in Arua taken over by Bee Natural Uganda Ltd. and reopened with new management

[1] Compiled from Enzama (2008) and interviews in 2010.
[2] NAADS = National Agricultural Advisory Services; UNIDO = United Nations Industrial Development Organization.

Beekeeping has since long been a traditional subsistence activity, especially in the highlands of West Nile. The activity was still rudimentary and largely unexploited as a farm enterprise. Mostly it took place in the form of gathering wild honey from caves, trees and anthills as part of collective socio-cultural activities of many communities in the region. Local self-made beehives were small and made of one piece and were often poorly sited. Wild honey gathering consisted of burning the natural colony if one suspected there to be sufficient honey, something, which disrupts or may destroy the bee colony. The honey then often contained traces of burnt materials and the smell of burnt bees and ashes. Honey was extracted from the comb by squeezing it with a cloth, without control of moisture. Others boiled the comb, altering the chemical composition of the honey. Honey was kept in small sized containers and cans. As a result productivity and quality of honey and beeswax was low and attracted low prices.

But beekeeping had considerable potential to introduce new technology, raise productivity in beekeeping and improve the quality of the honey. Firstly, there was considerable local and external demand for honey; secondly, the investment and operational costs of beekeeping is relatively low. It can thrive on marginal and infertile land that cannot support crop cultivation, as long as foliage is available. Other inputs (protective gear and equipment) can be shared with other beekeepers. As upfront costs are low, a farmer can break even within a year if good management practices are adopted. Beekeeping has limited vulnerability to disease and natural calamities; moreover it improves crop pollination and is an environmentally sound investment. Last but not least, it is not a physically strenuous exercise and youth as well as elderly, men as well as women can do it. It is a part time and seasonal activity where harvesting takes place twice a year.

When NAADS canvassed the selected sub-counties in 2001, there were an estimated 1,000 households scattered throughout the region regularly undertaking beekeeping activities. The formation and capacity building of farmer groups was the first challenge of institutional development. This was contracted out to NGOs. In 2001 the district government of Arua, NAADS and two NGO signed a memorandum of understanding for the purpose of formation and training of farmer groups.

Institutional change was needed and clearly associated with the introduction of new physical technologies of beekeeping, new farm level routines or practices. In addition new complementary institutional arrangements or 'social technologies' (Nelson, 2008) had to be designed to organise farmer groups and networks with other chain actors, with the purpose to impart knowledge and skills, jointly manage equipment, get access to micro-credit and eventually undertake group marketing.

The program soon ran into problems. Agricultural training institutions in Uganda did not offer course and expertise in beekeeping and related disciplines. In the whole of East Africa there was only one such institution in Tanzania. District entomologists stepped in to provide some technical assistance. NGOs had expertise and track records in capacity building in community groups and could engage farmers and provide capacity building in setting up and self-management of farmer groups, but were ill-equipped for specialist BDS in beekeeping.

Beekeeping is more developed in the relatively affluent central and western regions of Uganda. The Uganda National Apiculture Development Organization, a business interest association, draws most of its members from here. At that time, Mrs Maria Odido was its chairperson. She had taken a keen interest in the development of PMA and of NAADS and recognised their potentials. In 2002 she established a private limited company together with Mr Antonio Di Fonzo called Bee Natural Products Ltd. (BNP) in Kampala. In the same year she took samples of West Nile honey to an auction in the Netherlands where it was rated second to Brazilian honey in terms of quality.

The demand perspective for honey was generally considered rather positive both domestically and abroad. West Nile produces organic honey, which could potentially penetrate the fast growing demand for organic honey in Europe (Loon and Koekoek, 2006; cited by Enzama, 2008). In 2005 honey from Uganda was certified for export to the EU. Entering this market offered huge opportunities but also enormous challenges. After all, it implied developing a substantial agro-export chain, which would have to handle considerable volumes in order to become sustainable.

So, if demand conditions were favourable, a chain coordinator was needed that would be capable to organise this new export chain. In 2002, BNP Ltd. made a first step in this direction by setting up a honey processing plant in West Nile, located in Arua. This plant had an installed annual capacity of 600 metric tons. BNP aimed to produce honey and beeswax finished and labelled to international standards and ready for the consumer market. But setting up a continuous supply chain of quality organic honey for a plant of 600 metric tons is quite something else than creating farmer groups to be endowed with new technology and skills to produce better quality honey.

In October 2003 BNP and NAADS signed an agreement whereby BNP would assist in implementing the action plans of beekeeping farmer groups. The goal of the agreement was to speed up the adoption of improved technology by beekeepers to increase honey productivity and sales so as to diversify sources of household cash income. At the same time, the agreement served BNP to create its own network of suppliers of comb honey. Concretely BNP was to: (1) facilitate formation and strengthening of beekeepers associations (of beekeeper groups) for organised production and collective access to inputs and product markets; (2) offer extension services for commercial beekeeping; (3) introduce new technologies and beekeeping practices, notably better yielding and long lasting bee hives and harvesting gear and equipment; and (4) to buy the comb honey produced while beekeepers would reciprocate by selling their comb honey to BNP.

The first beneficiaries were 42 beekeeper groups in Arua. Groups in other districts were later incorporated into the agreement, covering about 5 groups per sub-county out of an average of 20 groups. The cost of the contract was equivalent to € 40,890.00 which was meant to cover costs of training artisans to make 'improved technology' beehives, extension service, distribution of beehives and demonstrations. This fund was managed by the NAADS secretariat.

The formation of beekeeper groups (as a social technology) was for NAADS primarily motivated by the need to reduce the costs of transferring knowledge about improved technologies and associated farm practices within the desired framework of a market for agricultural BDS. For BNP, however, the beekeeper groups were critical for the development of its own supply chain, ensuring continuous high quality supplies of honeycomb and reducing logistic and transaction costs. By investing in the relations with beekeeping groups and assisting NAADS with tasks, BNP expected to build up trust with beekeepers for future buyer-supplier relations.

Where NGOs were limited to organise farmer groups, BNP could continue imparting skills and technology, and train beekeeping groups in joint marketing, joint management of equipment and of demonstration sites. The formation of associations of beekeeping groups was important for BNP. This higher level of self-organisation of beekeepers would reduce the complexity of supply chain management for BNP. This however turned out to be much more difficult than originally foreseen. Joint sales managed by the associations would demand a more complex and transparent management system of accounting for group and individual contributions. Furthermore, associations would also become a stronger party negotiating contracts with BNP. Last but not least the associations required a high level of trust among farmers and farmer groups.

The improved technology centred on the use of a new type of beehive. Under the agreement, the Kenyan Top Bar (KTB) hive was introduced and popularised in the region to replace the traditional log hives. BNP set up two apiary technology demonstration and trial sites in each of the 12 participating sub-counties to demonstrate the use of the KTB hive for improved production. BNP introduced to the use of smokers for harvesting. Producers now also wear protective gears, gloves, and gumboots during harvesting. In this way, not only the quality of honey is improved but also the quantity that is harvested from one hive rises.

At the time of the agreement in 2003, the beehives were imported from Kenya and brought in from Kampala. This was not only time consuming but also costly. At the initiative of BNP, five artisans were trained and equipped by BNP with tools and machinery to produce hives locally. As a result, in Arua alone, the trained artisans have established three workshops and employ over 30 workers. The region now no longer imports hives. Business for these workshops is set to improve with the neighbouring Sudan and Congo placing orders for hives from West Nile.

Much as the high price is conducive for the young artisans to increase their earnings from the sale of hives, many smallholders cannot afford hives. Micro-credit is still hard to access for agriculture and related activities from financial institutions. In order to cope with this financial market failure, BNP started making agreements with individual farmers to distribute beehives to them and deducting the cost from the payment for honey over a period of two to three years.

BNP advisory services involved imparting apiary management techniques and production knowledge: how to locate good apiary sites, baiting of bees and techniques of determining the readiness of the honey for harvest, how to maintain apiary site to avoid infections and threats of ants, lizards and snakes, etc. The field officers of BNP paid regular visits to the apiaries to demonstrate the skills learned in theory and for purpose of comb inspection and quality checks. In this way, BNP was able to trace and control the quality of the production process right from apiary to the factory.

A common problem in beekeeping in West Nile was the low colonisation rate of beehives, which stood at only 60%. As more hives were being introduced, queen rearing became the answer to raise the colonisation rate. In 2007 the second agreement was signed between NAADs and BNP, which apart from up-scaling the previous activities now also included the setting up of demonstration sites for queen rearing. This agreement involved a total of € 73,293 of which 12.8% was an in-kind contribution from BNP as part of their normal interventions and as a lead firm of the export chain. The remainder came from the NAADS programme. This activity had difficulty getting off the ground due to the lack of BNP manpower with the requisite skills.

BNP constituted the primary link to the market: It was expected that BNP would buy all the honey from the producers. Likewise, producers were expected to sell their honey to BNP in exchange for the support offered in terms of inputs, technology, training and upgrading of products. A win-win situation was envisaged: the farmers would benefit from a ready and predictable market for their honey, reduced transaction costs, while BNP increased its assured sources of quality comb honey. This complex institutional arrangement can be classified as an informal relational contract.

Initially, farmers were happy and expectant with this arrangement because BNP was seen as a credible registered firm, with a location in the region, recognised by government and providing opportunities for acquisition of new and improved technologies and practices. However, the beekeepers were not party in the agreement between NAADS and BNP. Moreover, the agreement was silent on quality standards, price and delivery arrangements. The grading of the quality of honey was set and done by BNP at the factory and in absence of the producers. There were basically two grades of honey (A and B) according to moisture content, colour and scent. Furthermore, BNP tended to offer lower prices than other traders, which is something the company justified on the grounds of the subsidies in kind in the form of technical assistance, and implied financial costs of providing beehives with deferred re-payment.

Initially BNP paid promptly, but as the supply chain became more extensive and complex, payments delayed two weeks or more, while other travelling traders paid cash on delivery. Side selling by beekeepers increased as more traders visited Arua, attracted by its growing supply base. As side selling increased it became uneconomical for BNP to send out its truck to collect smaller quantities of honey from distant locations and effectively the spatial range of the supply chain shrank and with it the production supply base. By 2007 the factory was operating at not more than 25% of its capacity. Overhead costs rose, reducing further the ability of BNP to raise producer prices.

Clearly non-core activities had started to overwhelm BNP. Not only the number of farmer groups embracing apiculture rose rapidly, but also the number of activities undertaken by BNP to support the expansion and deepening of the honey supply base increased: from technical advice and training of beekeepers, co-financing the local production of beehives and their distribution, queen rearing and setting up the supply chain for processing and sale of honey and beeswax in domestic and export markets. With only two extension staff the quality of its service started to decline and eventually stopped in 2005.

The company was also hit by high turnover of staff. The employees complained of poor pay and terms of service and the director of BNP accused the employees of cheating the company and not accounting for some company funds. Employees who were fired became rival traders in honey. Consistency and continuity of service delivery to the beekeepers was thus undermined. The trust which BNP had started to build up with beekeepers, rapidly eroded. In 2008 the company decided to close down the factory as processed volumes had become uneconomical. Later on, in 2009 the processing plant in Arua was re-opened under new ownership and management and it is set to re-develop supplier relationships and regain beekeeper's loyalty.

What were the effects of the disruption of the BNP value chain on beekeeping in West Nile in 2008? Clearly BNP could not continue to remain a de-facto monopsonist controlling the entire value chain. Beekeepers became disloyal as evidenced by rapid rise of side selling and by BNP's inability to enforce the informal institutional arrangements with local beekeepers groups. Even beekeepers, who had obtained beehives from BNP were side selling and claimed that they could not deliver honey to BNP and thereby repay their loans in kind.

Clearly, BNP suffered from own management problems as evidenced by a lack of effective costing and contracting of non-core activities with beekeepers (and with which it had no prior experiences), lack of transparent quality assessment and pricing and the inability to manage the expanding supply chain as evidenced by increasing delays in payments and declining levels of service delivery. Supplier loyalty was high at initial stages as the benefits in the form of free services were visible and highly appreciated but declined later on as prices became a contentious issue and switching to the alternative option of side selling became easier to realise as more traders visited the area.

Furthermore, other actors introduced different technologies and alternative institutional arrangements to the region. Firstly, United Nations Industrial Development Organisation (UNIDO) developed a project in West Nile to promote beekeeping, small-scale processing of honey and marketing under arm's length market arrangements for ex-combatants of defunct rebel groups. Secondly SNV, a Dutch development organisation, which focused on larger beekeepers associations and provided organisational capacity building and facilitated links with credible (but mostly non-profit) organisations providing tailor made technical, logistical, market and financial support to expand the supported association's operations. SNV has supported 127 groups in the neighbouring Moyo District and the Netherlands Embassy subsidised beehives and equipment. Thirdly, enterprising beekeepers started investing in forward integration at a small scale. By starting their own small scale processing units, they undercut BNP while benefitting from the development of the improved technologies and production services among beekeepers. One example is the company 'Bee for Life' in the nearby Yumbe district, which buys honey from some 500 beekeepers.

As BNP no longer met the advisory and technology needs of the growing number of beekeepers, the local governments in the region through the NAADS started to expand contracts to private service providers to offer supplementary advisory and technology development services to fill the gap. By end of 2004, graduates from Nyabea Training Centre in Masindi, with elementary certificates in apiculture expanded the supply of advisory services in the region. By 2007 some 37 of these graduates were awarded contracts to support the beekeeper groups with advisory services and technology development. The dependence on BNP as a provider of key technological services declined.

What have been the overall local development results? The sector has experienced steady increase in the number of beekeepers from about 1,000 in 2002, before the agreement, to 4,000 in 2005 and over 6,300 in 2007. Idle resources like land, unsuitable for crop cultivation, has been put to use as apiary sites. Youth, majority of whom do not own and control land, is able to participate in the industry and also elderly members of the community who were unable to engage in crop agriculture could stay active in apiculture. The industry has created dynamism in the West Nile economy. Backwards and forward linkages have been developed. The local artisans who are making beehives and tailors who make protective wears have created opportunities for raw material input dealers. Additional processors set up operations and new traders arrived. BNP has also extended the honey chain to global markets, albeit temporarily. Beekeepers have gone back to the domestic arms-length market relations due to attractive domestic prices, incomplete contracts and lack of trust.

9.2.3 Institutional change for smallholder participation in export-agriculture in Northern Peru

The origins of our case can be traced back a Jesuit priest Jose de Bernardi, who developed the ideas concerning the creation of the Centre for Transfer of Technologies to University graduates (CTTU). Table 9.2 presents a timeline of the main events of CTTU and its activities (Helmsing, 2009). The CTTU targeted graduates of the regional universities and provided them with the opportunity to become young entrepreneurs forming their own agro-enterprise, dedicated to the growing of a high value export crop – asparagus. The intervention logic was primarily justified on political grounds: how to prevent that frustrated university graduates join the terrorist movements and instead of 'promoters of violence' become 'promoters of peace'. However, its application was primarily economic: how to form entrepreneurs and incubate their enterprises.

Crucial in the germination of his ideas was a chance encounter in 1990 between De Bernardi and the owner/manager of an innovative firm, TAL S.A. In 1991 the CTTU was created. Additional resources were subsequently obtained from a Dutch co-financing agency, which enabled it to expand staff and actually start its activities. A first land holding was acquired on desert land to begin incubation of agro-enterprises using advanced technology and farm management methods, developed with the 'foster enterprise' (TAL S.A.). Drip irrigation technology, which was new to Peru, required a minimum scale of operation in order to be economically viable. Single person enterprises envisaged as part of the incubation process were too small. This problem could be addressed by creating a cooperative in which irrigation assets would be pooled. But since the collapse of the agrarian reform, cooperatives had a 'bad reputation'. Therefore CTTU chose another collective solution well-known to its staff

Table 9.2. Timeline of events in La Libertad, Peru.[1]

1990	Principal protagonist meets leading entrepreneur: a learning alliance is formed
1991	Creation of Centre for Transfer of Technology to University (CTTU) graduates
1993	CEBEMO (later named CordAid) approves the CTTU project for funding (1993-1995)
1993	Acquisition of the land holding 'San Juan' from the CHAVIMOCHIC project (25 ha)
1993-1995	Promotion of first cohort of (12) young entrepreneurs – Drip Irrigation Production Unit (DIPU) 'San Juan', Moche
1994	CTTU itself is legally constituted as 'non-profit socio-cultural association'
1995	First promotion of Gravity Irrigation Producer Association (GIPA) groups (5) in Chao and Virú
1995	Second promotion of (10) young entrepreneurs – DIPU 'San Martin', Moche
1996	CEBEMO agreed to finance a second project phase (1996-1998)
1996	Agreement with community of Paijan – CTTU acquires 100 ha of communally held desert land
1997	Third promotion of (12) young entrepreneurs – DIPU 'San Jose', Paijan
1997	CTTU creates an agricultural enterprise 'Casuarinas', Moche
1998-1999	Serious damages by heavy rains caused by 'El Niño' and a drop in asparagus yields
1999	CordAid agreed to finance a third project phase (1999-2001)
1999	CTTU starts a parallel integrated local development project in Paijan, financed by Action Aid
1999	Fourth promotion of (10) young entrepreneurs, DIPU 'San Ignacio de Loyola', Paijan
1999	DIPU San Juan creates a Limited Company called 'Agro San Juan SAC'
1999-2000	Four new GIPA groups are formed in Paijan; two GIPAs formed 1988 close down
2000	Asparagus price drops in the international market (price war initiated by China)
2001	Formation of six new GIPA groups in Paijan; two GIPAs of 2000 close down
2000-2001	Export boom becomes a bust: falling international prices
2001	Drop in GIPAs (2) and GIPA membership (95) in Virú, Chao and Chimbote
2001	Government policy to create a plural and competitive system of agricultural BDS
2003	Governing board CTTU: CTTU to withdraw from credit operations
2003	DIPU San Jose, Paijan creates limited company called 'Agro Lider SAP SAC'
2004	CTTU secures two government funded projects providing business development services and chain coordination to contract farmers
2005-2006	New model replicated among small producers and companies
2007	Financial institutions accept CTTU model for group loans for export agriculture finance scheme

[1] Sources: field interviews and internal documents CTTU.

and existing in Peruvian civil law: a non-profit 'welfare organisation'. This became the Drip Irrigation Production Unit (DIPU), which legally owned the irrigation infrastructure, serving 15 to 25 one-hectare single person enterprises. The first DIPU started operations in 1995 with 12 university graduates after a long struggle to locate underground water for irrigation. It achieved spectacular yields, much higher than large-scale agro-companies in the region.

As the fame of the project spread to peasant communities in the valleys of the Department, these began to pressure the CTTU 'not to abandon them'. Resource conditions on peasant smallholdings, however, did not permit drip irrigation and technological constraints led to a complementary institutional adaptation. The Gravity Irrigation Producers Association (GIPA) was a new institutional arrangement created in 1995. This institutional arrangement does not own any assets but organises selected groups of young rural higher education graduates for learning, input distribution and group marketing. In the second half of the nineties a number of GIPAs were created in the valleys of Chao, Virú and Chimbote.

A major breakthrough in addressing the land constraint to incubate advanced technology enterprises (DIPUs) was the 'acquisition' of 100 hectares of desert land owned by the community of Paijan (in the North of the Department of La Libertad) on the condition that CTTU would stimulate enterprise development in DIPUs and GIPAs in the Paijan district.

Faced with market failure in the credit market, credit then became the most binding constraint. CTTU addressed this by assuming responsibility for a large loan obtained from the Canadian Counter Value Fund. The number of DIPUs and especially GIPAs increased rapidly. In order to serve the new enterprises with high yielding varieties, CTTU set up its own nursery in 1997 where it produces seedlings under controlled conditions. Seedlings were provided as a service, free of charge, to DIPU and GIPA members.

The asparagus export boom in the region received a big stimulus with completion of the CHAVIMOCHIC project. The state sold by public tender 9,000 hectares of dessert land alongside a newly constructed irrigation channel to large companies, many of which invested directly in asparagus production. As a result, the region became a leading exporter. The asparagus boom attracted also related and supporting industries and service providers (input distributors, sale and hire of farm equipment, etc.) as well as rural labour that migrated from the peasant community highlands of the Andes to work the fields in this coastal region. Within a period of five years a new regional export base developed around one single crop, asparagus.

In 1999 CTTU applied again to the Dutch co-financing agency for financial support for a third period. This third application was also successful. The number of applicants to CTTU grew, attracted by the high incomes earned in the export activities. The relative resource abundance of CTTU in those years resulted in less strict selection by CTTU of potential entrepreneurs.

In 1998-1999 the 'El Niño' phenomenon struck, causing heavy rains and flooding in the valleys, resulting in damages to irrigated fields. Crop yields declined in GIPAs in these valleys, but DIPUs situated in the desert were not affected. Before small farmers could recoup their losses, an aggressive Chinese export drive in the world market led in 2000-2001 to a fall in asparagus world prices. The number of GIPAs as well as GIPA membership declined rapidly. Members defaulted on their loans and left the CTTU with their accumulated debts. The growth of DIPUs also stagnated as economic prospects had declined.

The institutional change agent, CTTU, suddenly found itself in crisis. Thanks to 'bridging finance from Cordaid, CTTU could engage in extensive consultations with its principal stakeholders and producer groups. The new institutional choices made by CTTU were influenced by three main factors: (1) important changes in the broader institutional environment, notably a new agricultural policy of the Peruvian government, which created new opportunities for CTTU based on its acquired reputation; (2) the response of Peruvian export firms to the competitive challenge of China; and (3) the vision of the CTTU about its own future role.

The World Bank sponsored INCAGRO project signified an important change in Peru's agricultural policy. Its aim was to create a market for business development services for commercial agriculture, whereby private firms and NGOs provide extension services to groups of producers, which were co-financed by the Government. At the same time the Peruvian Government created so-called 'second tier' funds to finance the expansion of commercial agriculture. The INCAGRO project was a national scheme to which (independent) groups of agro-producers could apply often in association with agro-industry. Thanks to its accumulated experience and reputation, CTTU could organise groups of producers and team up with agro-industrial firms and could make several successful bids.

The second factor refers to technological innovations in transport of horticulture exports in neighbouring Chile and their adaptation to the Peruvian asparagus (the use of air controlled containers extending the fresh life of horticulture products by changing the percentage of oxygen) and the extension of this shipping service to Trujillo, Peru. Thanks to these technological innovations, the Peruvian firms succeeded to redefine their market niche by switching from preserved white asparagus to fresh green asparagus. Since then they have become world leader in fresh asparagus, leaving China to dominate the world market of preserved (canned) asparagus.

The technological innovation in production and logistics related to asparagus also made it easy to adapt these to other high value export crops. This made it possible for Peruvian firms to diversify their export crops. The new export products were annual crops (artichoke, peppers, etc.) reducing the high risks associated with investment in semi-perennials, such as asparagus. Based on prior successful collaboration with CTTU, export companies shared their learning experiences and on that basis CTTU could relatively quickly adapt training packages to the new crops.

The Government policy of financing commercial agro-export set the terms for the new contractual arrangements. Banks would provide credit against the presentation of a contract with an agro-export firm. This led to a local adaptation of a contract farming model: CTTU became a chain coordinator, providing chain coordination, agricultural extension and related business services to small producers. For small producers the CTTU provided 'transaction opportunities' in markets not accessible to them individually, notably in the markets for export crops, for inputs, and for credit. The risks of operating in volatile export markets were managed by means of new interlocking contractual arrangements between CTTU and smallholder, between smallholders and export companies and between smallholders and banks with CTTU as co-signatory. In this new institutional set up, CTTU became a non-profit or social enterprise with a mission to serve small producers but charging for its services.

How successful has the CTTU been in its original objectives? The results are mixed: the original plan of incubation of individual enterprises in combination with a desert land colonisation scheme, based on the DIPU model was, in the end, not successful. The chosen institutional arrangement 'froze' the incubation process. The institutional arrangements of a welfare organisation did not permit unsuccessful members to exit with compensation for their past efforts. CTTU has been most successful with the institutional model it had neither initially designed nor foreseen. This was the new GIPA model (a combination of collective action, contract farming and social enterprise services), creating 'transaction opportunities' for educated children of the 'parceleros of the agrarian land reform' and incubating their new enterprises on small plots on former irrigated estates.

CTTU's success can be explained by its capacity to adapt to changing circumstances, aided by long term funding from a 'patient' donor, a learning alliance with an innovative private firm and by having built up a reputation with large export companies. The very transformation of the regional economy created 'a critical mass', economically and politically. Economically, in so far that the geographical concentration of asparagus production attracted specialised suppliers and services to the region from which also small producers benefitted and because large firms were able to respond successfully to the competitive challenge of China. In political terms in the sense that export business leaders from the region were invited to help give shape to the new agricultural policies of the Government (INCAGRO project) and because large companies were able to lobby the Government for infrastructural improvements (roads and sea- and airport).

Since its re-engineering, the CTTU has selected more than 420 persons for its entrepreneurship and enterprise development programme. Table 9.3 gives an overview of the status in 2008 of all incubatees since 2000. Of these, nearly 10% in that year were participating in the programme. 57 entrepreneurs were engaged in export chains coordinated by CTTU and another 60 were doing so independently from CTTU. Another 30 persons could find employment in the same sector, thanks to the competences acquired through CTTU. In almost 100 instances, CTTU was unsuccessful. The incubatees, after some time, switched back to traditional crop cultivation and farming practices. There were also three categories of what we could characterise as unsuccessful instances as people moved out of agriculture altogether, either they migrated, switched to non-agricultural occupations or social reasons explained their exit. Furthermore, a significant group of persons (11%) existed without any information on their whereabouts.

Table 9.3. Status in 2008 of the Centre for Transfer of Technology to University (CTTU) incubatees since 2000.[1]

Status	2008	%
Currently in process of incubation	41	9.7
Start-up enterprise operating in chains coordinated by CTTU	57	13.5
Start-up enterprises operating independently in group based agro-export cultivation	60	14.2
Employed in the agro-export sector	30	7.1
Unsuccessful incubation – returned to traditional cultivation	98	23.2
Unsuccessful incubation – moved into non-agricultural employment	39	9.2
Unsuccessful incubation – rural to urban migration	14	3.3
Unsuccessful incubation – due to social reasons (incl. health)	28	6.6
Other	9	2.1
Without information	47	11.1
Total	423	100.0

[1] Source: registers CTTU, Trujillo, Peru.

In order to conclude on the performance, Table 9.4 defines 'success' and 'failure' rates. Criterion 1 is the strictest definition of success: Have incubatees become independent entrepreneurs who now operate their enterprises in agro-export crops on their own or with independently formed groups? Using this criterion only 14% of the incubatees of CTTU can be considered successful. Criterion 2 recognises that small farm enterprises face systemic market failures and need 'allies' who help overcome these. The CTTU performs this role through its coordination of the agricultural production segment of agro-export chains. In this case the success rate rises to 28%. Criterion 3 has the broadest success definition. For people who fail as entrepreneurs, but who remain employed within the agro-export chains, one cannot conclude that the investment has been a waste of resources. The investments continue to yield social benefits. In this case the 'success rate' of CTTU rises to 35%.

We can also look at the performance of CTTU looking at the failure side. As shown in Table 9.4, the aggregate failure rate is 42%. That is to say, 4 out of every 10 persons who participated in the CTTU programmes did not form agro-export enterprises or remained active in that sector. However, only 2 in 10 reverted back to traditional farming practices. The other 2 in 10 for various reasons left the agricultural sector completely. This is a general characteristic of rural processes of change. It would most likely have happened irrespective of the CTTU intervention. In that sense it is a kind of dead weight factor that needs correction in the evaluation of the results. Taking this into account it can be concluded that the overall CTTU performance can be considered positive indeed.

Table 9.4. Success and failure of the Centre for Transfer of Technology to University (CTTU) driven change process.

	2008	%
Success		
Criterion 1: independent entrepreneur (without any assistance from CTTU)	60	14.2
Criterion 2: independent entrepreneur (with or without assistance from CTTU)	117	27.7
Criterion 3: active in agro-export chains + agricultural employment	147	34.8
Failure		
Unsuccessful incubation – return to traditional cultivation	98	23.2
Unsuccessful incubation – other employment (non-agricultural)	39	9.2
Unsuccessful incubation – rural to urban migration	14	3.3
Unsuccessful incubation – social factors	28	6.6
Aggregate rate	179	42.3

9.3 Reflecting on institutional co-innovation

What are the main commonalities among and the main differences between the two cases of co-innovation? There are at least four common elements and five main differences in the process of institutional co-innovation.

First of all, the key institution itself consisted of the rules concerning the formation of similar sized farmer groups (in practice 10-15 members). In both instances the main motivation was to achieve economies of scale in capacity building, imparting new skills and practices around a new physical technology and farm level practices. The social technology of forming groups, for group-based technical assistance, learning and experimentation, was complemented by use for group level management of joint assets (experimental stations, irrigation) and joint marketing. In both instances, change agents had to overcome market failure in input markets (notably finance, but also key inputs such as queen bees and quality seeds) in order to enable small farmers to enter the new product market. Both BNP and CTTU undertook micro finance lending to enable small farmers to acquire loans to finance the new activity. In both instances, the innovative agent nearly collapsed under this weight of these 'non-core' activities, which aimed to eliminate critical and binding resource constraints of small farmers.

Secondly, in both instances the initial institution – the farmers group for organising and managing technical assistance – has been adapted to suit the coordination needs of other economic agents. In Uganda, BNP Ltd. transformed the farmer groups into a full-fledged supply chain institution to be able to export quality honey; in Peru the institution was adapted (CTTU ceased forming DIPUs and redefined the GIPAs into a more flexible annual contract farming group operating in export chains).

Thirdly, in both instances the new institution was interlinked with other parallel institutions in order to export high value horticultural crops: in Uganda, there were bi-lateral interlocking contracts between BNP and contact farmers for purpose of supply chain logistics and with beekeepers on distribution of new beehives and on sale of honey; in Peru, the GIPA was the basis for multi-lateral interlocking contracts between small farmer groups – CTTU on the one hand and agro-export firms and banks on the other hand. In both cases products quality standards were key to access international markets. The institutional rules on standards were neither negotiated nor independently verified but in both cases defined upfront by the dominant buyer(s) of the product: in Uganda BNP itself and in Peru the agro-export firms. In Peru there were no collective assets involved in the GIPA. In practice, beekeeper groups in West Nile also did not hold collective assets.

Fourthly, in both instances there were important power asymmetries, which influenced the direction of institutional co-innovation. In Uganda BNP was initially a monopsonist and effectively tried to turn the beekeeping groups into larger beekeeper associations as key nodes in its supply chain; in Peru the agro-export firms were much more powerful than the farmer groups as the former were vertically integrated processing firms, producing high value crops on their own large-scale farms. But in both instances the market situation strongly influences the degree to which power holders could exploit their power advantage. In Uganda, the growth of the industry attracted new rival traders and some beekeepers moved forward to expand in comb honey processing, thereby opening up new market outlets for other beekeepers, undermining the monopoly of BNP and its associated supply chain institutions. Arm's length market arrangements gradually replaced value chain based networks. In Peru the agro-export firms continued to dominate the export chains, but in global markets they were also price takers.

Having enumerated the main commonalities, let us now look at the five main differences between the two cases. First of all, the products for which new physical and social technologies needed to be developed were quite different. Asparagus growing is a full time and perennial crop, while beekeeping is a part-time seasonal activity. This has important implications for the intensity with which agents 'live' by the new institutional arrangements. In the one case these concern the primary occupation of the small farmer, while for the other they relate to a complementary seasonal and part-time activity. The relative importance of the income derived from honey was much smaller than the income generated by asparagus growing and so were the risks. Not honouring commitments with BNP had less social consequences within the beekeeper communities. In Peru poorly performing small farmers were excluded from farmer groups in subsequent rounds. Upfront entry barriers associated with the product also differed considerably. Asparagus growing required minimally one hectare of irrigated land and access to water (tube-well) and a substantial amount of working capital; in beekeeping there was hardly any entry barrier and it required relatively small cash outlays. Asparagus growing was new for most small farmers, while beekeeping was a traditional practice among many communities in West Nile.

Secondly, there were important differences in the selection process of small farmers. In Uganda's practice any rural household could participate; only large-scale farmers were excluded. In Uganda, the selection process was primarily bureaucratic: the selection of the sub-county in which NAADS would operate. In Peru there was a clear and upfront selection of young and educated small farmers (initially urban professional university graduates, later rural young farmers and graduates of agro-technical institutes). The scale of the process was quite different: in Peru the total number of small scale farmers was less than 500 in 2008; in Uganda it involved an increase from 1,000 in 2001 to 6,500 beekeepers in 2008; in Peru the same farm level technology and the associated social technology of small farmer groups was replicated to other high value export crops (such as paprika, peppers, etc.). This was not the case in Uganda.

Thirdly, and as regards actors, there were also important differences. In Uganda, the NGOs played only a supportive role. The governmental NAADS programme was the principal driver. A commercial firm (BNP Ltd.) played a key role in creating and transforming the honey value chain adapting and thereby extending the original institutional arrangements. The company initially assumed a social entrepreneurial catalytic role but later backtracked as financial and human resource implications overwhelmed it, to continue as a purely commercial firm. In Peru the NGO (CTTU) was the principal catalyst, which had a commercial firm as its ally; in the process the NGO adapted itself to changing financial circumstances (the termination of the project subsidy from a Northern NGO donor) and became a non-profit but market oriented NGO or social enterprise.

Fourthly, government policy in Uganda, and specifically the NAADS programme, played a fundamental role in shaping the institutional co-innovation. These institutional innovations were centrally decided, designed ex-ante and in detail by public officials with strong influence of the World Bank and with the aim to create market for technical assistance and to reduce the cost of implementing the new policy. Complementary innovations were made locally, by trial and error and in a decentralised manner involving other economic agents (BNP). Later on in the process, again other agents created locally rival institutional innovations (UNIDO, SNV) and rival traders offered arm's length contracts, thereby increasing the choice for small farmers but also undermining the role of beekeeper groups in the production and marketing process. In Peru, the CTTU initiated an institutional design process in a manner of trial and error, strongly guided by the unfolding technology of irrigated high value export crops growing and adapting its own designs along the way, discontinuing the DIPU and developing new and complementary institutional arrangements as the policy environment changed favourably with the new government programme of market based technical assistance (in the form of the World Bank financed INCAGRO program) and complementary financial policy (2nd tier funds for innovative SMEs and commercial small export farming). The Peruvian Government policy was not leading but provided an important tipping point in the evolutionary change process led by CTTU.

Fifthly, there are interesting differences as regard the direction of the process. In Peru and much in contrast to established NGO doctrine and practices: economic empowerment was considered fundamental to be achieved first, and political empowerment of disenfranchised groups would come later as an outcome. It should be said that in reality no evidence of the latter was found. In Uganda, the political empowerment of small farmers in order to become a stronger market party was a central feature of the NAADS policy. Without farmer groups, small farmers would not able to become an active player in the market for technical assistance and engage private sector (and NGO) suppliers of extension services. In practice however, political empowerment also followed economic empowerment: a number of successful men and women active in beekeeping in West Nile became sub-county and district level political leaders. In Peru the process was initially heavily supply driven: the unfolding technology strongly influenced the institutional options that were conceived (the DIPU). But local social demands made CTTU to adapt its institutional arrangements by creating the GIPA. Later on the new policy environment and the market demanded greater flexibility and the GIPA was adapted and farmer groups were formed more flexibly around product specific interlinked contracts. In Uganda, institutional adaptations also occurred: the small farmer groups formed to change the rules of technical assistance were transformed into supply chain nodes with their own specific institutional arrangements. However, the growth of the industry instead of consolidating the new institutional arrangements, led to the arrival of rival institutional agents and alternative options and in the end to the demise of the BNP institutional monopoly. Side selling played a key role: in Uganda BNP contracts were informal and entry of rival agents was much easier. This stood in sharp contrast with Peru where side selling was much more difficult due to the formal contracts and considerable entry barriers in processing and exporting.

Finally, let us consider the questions formulated at the beginning of this paper. We find in both instances strong evidence in favour of institutional co-innovation in different aspects of value chain development (in physical and social technologies, organisational innovations for learning, joint marketing, contract for quality products between chain actors, with innovations in complementary markets for inputs and finance). There is in both instances 'nestedness' of institutions whereby national public institutions give room for private institutions at the level of the chain (respectively level 2 institutions and level 3 institutions as defined by Williamson). However this 'nestedness' is not always present as is shown by both cases. In Uganda the process was top down and preceded value chain development, while in Peru there was a certain degree of bottom up creation of 'nestedness' as leading entrepreneurs demanded government to support the new export agriculture with institutional innovations in extension and finance. These new public policies followed and came for CTTU at the right time.

Is there a process of 'climbing the institutional ladder' as formulated by Brousseau and Raynaud suggested, whereby locally generated institutional innovations acquire greater acceptability? In the case of Peru this was certainly the case as CTTU initiated institutional arrangements became accepted by agro-export firms and banks, but competition between different institutional agents may cause a fall from the institutional ladder, as happened with the institutional arrangements instituted by BNP in Uganda.

While specific innovations may be designed upfront, co-innovation of complementary institutional innovations is rarely fully designed upfront but co-innovation implies a certain degree of decentralised experimentation. There may be several reasons for this, resource constraints among the principal actors (BNP and CTTU) in the face of market failures arising from the behaviour of other agents (e.g. in credit markets). But in my view the most important constraint on designed co-innovation is 'situated bounded rationality'. Rarely the principal innovation protagonists can foresee all contingencies and considerable 'on the ground' experimentation is needed to ensure that all complementary and co-evolving innovations match.

References

Benin, S., Nkonya, E., Okecho, G., Pender, J., Nahdy, S., Mugarura, S., Kato, E. and Kayobyo, G., 2007. Assessing the impact of the National Agricultural Advisory Services (NAADS) on the Uganda rural livelihoods. IFPRI Working paper 00724, International Food Policy Research Institute, Washington, DC, USA.

Brousseau, F. and Raynaud, E., 2007. Climbing the hierarchical ladders of rules: the dynamics of institutional framework. Working Paper, University of Paris, Paris, France.

Chang, H.J., 2002. Breaking the mould: an institutionalist political economy alternative to neo-liberal theory of the market and the state. Cambridge Journal of Economics 26: 539-559.

Chang, H.J., 2010. Institutions and economic development: theory, policy and history. Journal of Institutional Economics 7: 473-498.

Enzama, W., 2008. The quest for economic development in agrarian localities: lessons from West Nile, Uganda. Institute of Social Studies, The Hague, the Netherlands.

Evans, P., 1995. Embedded autonomy: states and industrial transformation. Princeton University press, Princeton, NJ, USA.

Gomez, G., 2008. Making markets. The institutional rise and decline of the Argentine Red de Trueque. PhD Thesis, Institute of Social Studies, The Hague, the Netherlands.

Helmsing, A.H.J., 2009. Smallholder participation in high value agro-export chains in Peru. A study of the co-evolution of technology and institutions. Paper presented at the Workshop on Globalisation and Changing Geographies of production and Innovation: Implications for Workers, Firms, Regions and Countries. November 5-7, 2009. Utrecht University, Utrecht, the Netherlands.

Hodgson, G.M., 1988. Economics and institutions. Polity Press, Cambridge, UK.

Hodgson, G.M., 2006. What are institutions? Journal of Economic Issues 40: 1-25.

Kingston, Chr. and Caballero, G., 2009. Comparing theories of institutional change. Journal of Institutional Economics 5: 151-180.

Maseland, R., 2011. How to make institutional economics better? Journal of Institutional Economics 7: 555-559.

Nelson, R.R., 2002. Bringing institutions into evolutionary growth theory. Journal of Economic Perspectives 12: 17-28.

Nelson, R.R., 2008. What enables rapid economic progress: what are needed institutions. Research Policy 37: 1-11.

Nelson, R.R. and Sampat, B.N., 2001. Making sense of institutions as a factor shaping economic performance. Journal of Economic Behaviour and Organization 44: 31-54.

North, D.C., 1990. Institutions, institutional change and economic performance. Cambridge University Press, Cambridge, UK.

North, D.C., 2005. Understanding the process of economic change. Princeton University Press, Princeton, NJ, USA.

Van Loon, M. and Koekoek, F.J., 2006. Export opportunities for African organic honey and beeswax. A survey of the markets in Germany, the United Kingdom and the Netherlands. EPOPA 2006, the Netherlands.

Williamson, O.E., 1975. Markets and hierarchies: analysis and anti-trust implications: a study in the economics of internal organization. Free Press, New York, NY, USA.

Williamson, O.E., 2000. The new institutional economics: taking stock, looking ahead. Journal of Economic Literature 38: 595-613.

10. Towards stable access to EU markets for the Beninese shrimp chain: quality, legal and marketing issues

S.A. Adekambi[1], D.S. Dabade[2], K. Kindji[3], H.M.W. den Besten[4], M. Faure[3], M.J.R. Nout[4], B. Sogbossi[5] and P.T.M. Ingenbleek[1]*

[1]Wageningen University, Marketing and Consumer Behaviur Group, Hollandseweg 1, 6706 KN Wageningen, the Netherlands; [2]Laboratoire de Biochimie Microbienne et de Biotechnologie Alimentaire, University of Abomey-Calavi, 01 B.P. 526 Cotonou, Benin; [3]Maastricht European Institute for Transnational Legal Reserach (METRO), Maastricht University, P.O. Box 616, 6200 MD Maastricht, the Netherlands; [4]Laboratory of Food Microbiology, Wageningen University, Bornse Weilanden 9, 6708 WG Wageningen, the Netherlands; [5]Université de Parakou, Faculté des sciences économiques et de gestion (FASEG), BP 123, Parakou, Benin; paul.ingenbleek@wur.nl

Abstract

Traditionally, the economy of Benin has strongly depended on a single crop, namely cotton. Since 2006, the Beninese government has aimed to diversify exports, in particular focussing on high-value export products such as shrimp. Stable market access for shrimps is, however, hindered by their microbiological and chemical characteristics which influence product quality and safety. In the international market, these quality aspects have legal implications, potentially leading to import bans if safety standards are not met. This chapter examines the quality and legal issues of the Beninese shrimp chain and discusses the responsiveness of the chain to these issues. Using an interdisciplinary analysis, the chapter draws preliminary conclusions on how a stable access of Beninese shrimps to the international market can be achieved.

Keywords: shrimp quality, legislation, market integration, institutions, value chains, Benin

10.1 Introduction

Connecting small-scale producers to export markets has become a popular development policy strategy to improve the livelihoods of small-scale agricultural producers (Gulati *et al.*, 2007). Traditionally, the export positions of many developing and emerging countries depend heavily on one or two export commodities, which makes them vulnerable to price shocks of those commodities on the world market. Diversification into high-value export products has therefore become an important policy to sustain growth and food security and to alleviate poverty (Weinberger and Lumpkin, 2007). Evidence from Asia has shown that such policies encourage innovation (e.g. Deshingkar *et al.*, 2003; Pokharel, 2003), enhance farmers' incomes and stimulate exports (e.g. Birthal *et al.*, 2007; Pingali and Khwaja, 2005).

However, creating access to the world market is a difficult process for any value chain. Prior research has shown that success is not guaranteed (e.g. Diaz Rios and Jaffee, 2008; Kambewa, 2007; Jongwanich, 2009). In fact, the ability to meet increasingly more stringent food safety standards on developed country markets has become a major challenge for developing country exports (Jongwanich, 2009). In this regard, food standards are significant impediments to trade for many developing countries (Henson and Loader, 2001), resulting in adverse effects on competitiveness, and hence on market access (Diaz Rios and Jaffee, 2008). Kambewa (2007) shows, for example, how the establishment of an export chain for Nile perch from Lake Victoria eventually deteriorated the livelihoods of fishermen and the ecological sustainability of the lake. The case is particularly interesting because fish is not only a high-value product as compared to traditional commodities, but also a vulnerable product that gets easily contaminated, especially in environments with ambient temperatures and the absence of resources and institutions that can secure its quality, like refrigerated transport and laboratories that can check safety and quality. Evidence from other, yet comparable, export chains would help to create a deeper understanding of the factors that should be taken into account when establishing export connections to high-income markets like the European Union (EU) or North America, for valuable, yet vulnerable food products.

This chapter analyses the efforts of the Beninese shrimp chain to connect to the EU market. Benin is a country in sub-Saharan West Africa. Poverty reduction is the key focus of its economic and social policies as 75% of the Beninese population earns less than 2 US dollars a day (World Bank, 2010). Agriculture is the most important economic sector in the country, not only because of the number of active people employed but also for its contribution to exports. Approximately 70% of the active population gain their revenue from agriculture, and this sector contributes up to 80% to export revenues (SCRP, 2007). Cotton constitutes the primary export commodity with about 40% of gross domestic product (GDP) and over 80% of official export receipts (Aregheore, 2009; PSRSA, 2010). The cotton sector is, however, vulnerable to price volatility due to, among others, trade interventions like foreign subsidies (Baffes, 2011), leading to foreign currency losses and livelihoods deterioration. In 2006, the Beninese government therefore aimed to stabilise its export position by increasing the exports of high-value products, including shrimps. Increasing shrimp exports is a valuable strategic choice considering the past importance this sector had in the economy of Benin. As the second export product after cotton, shrimps were essential for their beneficial effects in terms of employment, foreign currencies earnings, and tax revenues. However, the shrimp export sector has been in decline over the last ten years due to difficulties in complying with the requirements of the EU, its major market. Therefore, it appears crucial that development goals address issues related to such a promising sector, and attempt to restore the conditions that can contribute to its expansion.

This chapter aims at outlining the factors that have contributed to the collapse of the shrimp export sector, and attempts to identify the conditions that would enable a sustainable revival of the sector. In the following, this chapter first provides a short conceptual background and description of materials and methods. Next, it offers a brief description of the chain including a description of its history. This is followed by a discussion of the microbiological quality of shrimps, and a description of the institutional environment that surrounds the shrimp chain. The chapter continues with a discussion of the measures taken for improvement and the responsiveness of the shrimp sector to those measures. The chapter concludes with a discussion section and policy recommendations.

10.2 Conceptual approach

10.2.1 Market integration

Building on the definition of Maertens *et al.* (2011), export market integration (EMI) refers to the share of products sold to export markets. EMI is perceived in development literature as a way-out of poverty because it might facilitate access to production inputs, enhance improved technology adoption and hence increase farm income (e.g. Bernard *et al.* 2008; Maertens *et al.*, 2011). Yet, EMI remains a major challenge for governments and development organisations in developing countries. Numerous studies have been conducted to examine and deal with the issue. According to most of these studies, three main institutional constraints stand in the way of EMI of small-scale producers: high cost of doing business, such as transportation costs (e.g. Bougheas *et al.* 1999; Buys *et al.*, 2010), lack of access to investment capital like credit (e.g. Nieto *et al.*, 2007; Van Greuning *et al.*, 1998) and limited access to market information (e.g. Marter, 2005). High transaction costs, which are due to the lack of transport facilities and inadequate farm-to-market roads, limit small-scale producers' access to traders and basic market information (e.g. Marter, 2005). Lack of access to investment capital limits small-scale producers' capacity to invest in quality. Lack of access to basic market information reduces the ability of small-scale producers to take effective production decisions. The conclusion in the development economics literature is that these institutional constraints need further improvements to facilitate small-scale producers' EMI. This is particularly important for high-value products.

10.2.2 High value products

High-value agricultural products (HVAP) are non-staple agricultural products with a high return, including fruits, vegetables, flowers, animal products, and fish (CGIAR, 2005; Gulati *et al.*, 2007). Because of their high economic value, they represent an important counterweight to traditional export commodities, such as cotton, coffee, tea, sugar and cocoa. The past decade has witnessed increasing world demand for HVAP due to changing consumption preferences, in particular in developed countries. This demand has gone up by 6-7% per year (World Bank, 2010). Such a growth in demand for HVAP is explained by shifts in consumers' preferences, sustained rise in income and urbanisation (e.g. Birthal *et al.*, 2007; Gulati *et al.*, 2007).

Diversifying agricultural exports towards HVAP is argued to play a significant role in poverty alleviation, sustainable growth and food security (Birthal *et al.*, 2007; Maertens *et al.*, 2011). HVAP are suggested to support poor communities to make the transition from subsistence to market-oriented agriculture through, for instance, vertical linkages between small-scale producers and their buyers (Birthal *et al.*, 2007; CGIAR, 2005; Weinberger and Lumpkin, 2007) and create opportunities for small farmers to raise their income (Gulati *et al.*, 2005). For example, Maertens and Swinnen (2009) find that HAVP exports have significant and positive effects on poor household incomes in Senegal.

Shrimps constitute one of the HVAP identified in Benin's Growth and Poverty Reduction Strategy to diversify agricultural exports and alleviate poverty (SCRP, 2007). The sector has significant potential for exports as world-wide demands for shrimps are increasing by 3% a year (United State Trade Representative, 2005).

10.2.3 Food safety

Because HVAP are generally more vulnerable than staple products, they often put specific pressure on the institutions that surround the chain. Food safety and quality assurance therefore rank high on the policy agenda (e.g. Ababouch *et al.*, 2005). As a result, there has been an international drive towards reforming fish inspection systems to move away from end-product sampling and inspection into preventive Hazard Analysis Critical Control Point (HACCP)-based safety and quality systems (Ababouch *et al.*, 2005) to ensure that food does not cause harm to consumers. A food chain approach has been developed in which the responsibilities for food safety and quality is shared by all chain members (FAO, 2003). The chain approach calls for integration between chain members and disciplines (Ababouch, 2006). Logically, appropriate legislation must be backed-up by the institutions that ensure an effective implementation of the regulatory requirements.

In the specific case of shrimps, a key challenge is to ensure through microbiological and chemical surveillance that raw shrimps are free from contamination that could affect the safety of the processed product. In the same way, stakeholders must have appropriate skills to implement technical requirements deriving from the regulatory framework. In a nutshell, the development of a modern and competitive agro-food system involves *inter alia* the ability to detect or demonstrate the presence or the absence of biological, chemical or physical hazards, the verification or certification of traded products with respect to established food safety risks, undertaking scientific analysis of hazards, and establishing or maintaining systems for hygienic practices in food product handling and transformation (Jaffee and Henson, 2004).

10.2.4 Informal institutions

Companies in the formal sectors of the export chain, such as exporters, importers, processors, and retailers, are typically influenced by formal institutions like laws and regulations. Primary producers and traders, often constitute informal stages of the chain that are weakly influenced by formal institutions like laws and regulations (Burgess and Steenkamp, 2006; Castells and Portes, 1989). The weak influence of formal institutions brings informal institutions, like cultural norms and value, more to the forefront. One of the typical cultural characteristics of subsistence contexts is embeddedness (e.g. Burgess and Steenkamp, 2006). Embeddedness is referred to as a desirable relationship between an individual and a group that helps to maintain the *status quo* and that limits actions that might disrupt group solidarity or a traditional order (Licht *et al.*, 2005). The question arises how differences in cultural institutions influence the development of the appropriate behaviour that secures food quality and safety.

10.3 Materials and methods

Research can potentially help to accomplish the challenges described above by understanding the microbiological challenges of the product, the legislation that protects consumers from unsafe shrimps, and the marketing mechanisms in the chain that enable small-scale shrimp fishers to respond to these issues. This chapter describes and integrates recent insights from three on-going PhD studies in food quality and safety, international trade law and marketing, respectively. The chapter builds on the material of these studies, which include an analysis of the existing literature in the three domains and additional desk and field research. With respect to the quality and safety part, this chapter draws in particular on a survey conducted in the Beninese shrimp sector, involving 325 shrimp fishers, 128 intermediate traders (vendors and traders), 12 collectors, and 3 shrimp freezing plant managers to determine the conditions under which shrimps are handled at the fishing areas and in the supply chain towards the plants.

From the legal perspective, the chapter draws on analyses of the food laws in Benin and the EU. This desk research essentially studied the development in Benin law with regard to trends in EU law and assesses the potential interactions between those two sets of law. The desk research has been supported by five interviews with actors such as officers of the Directorate of Fisheries, shrimp projects managers and the quality manager of one processing plant.

The marketing part has been studied by means of desk research combined with interviews and personal observations. The desk research included an in-depth analysis of, on the one hand, publications in the marketing and development literature, and on the other hand, policy documents, research and project management reports in the domain of shrimp. We completed the desk research with interviews with five experts including researchers, shrimp project managers and fishing directorate agents.

10.4 The Beninese shrimp chain

Shrimps are caught by small-scale fishermen as an informal economic activity in the southern part of Benin. Shrimps are caught in waters that are connected to the sea, which is essential for their development. The life cycle of shrimp includes several phases divided between the sea and lagoons or lakes. Almost the entire production comes from three main lakes/lagoons in this region: Lake Nokoue, Lake Aheme, and lagoon of Porto-Novo.

The traditional chain is oriented towards domestic consumers and consumers in what can be called the regional market, consisting of neighbouring countries, such as Nigeria, Togo, and Ghana. As most of this trade is informal, there are no official statistics on the shrimp quantities that are exported to neighbouring countries. The chain is mainly controlled by traders who buy raw shrimps from shrimp fishers and/ or from vendors. Traders subsequently process the shrimps to increase their shelf life, by smoking, or sometimes by frying or drying them. Social relations exist between several chain actors. Some traders provide fishers or vendors with advance payments to maintain loyalty.

Figure 10.1 shows the Beninese shrimp chain. The sector provides work to approximately 21,000 permanent artisanal shrimp fishers (Délégation de la Commission Européenne au Bénin, 2009), of whom most are men (Le Ry *et al.*, 2007). By 2002, shrimps represented around 1% of Benin's exports (SCRP, 2007). From 1993 to 2003, shrimp exports generated substantial revenues and foreign exchange for the country (STDF, 2008). The export value of the shrimp sector increased from 850,000 Euro in 1999 to 3.3 million Euro in 2002.

10.4.1 Brief history

When the Beninese government recognised shrimp as one of the high-value products that could be exported to high-income markets, an export-oriented chain was set up mainly targeting the European Union (EU) where the country has historical connections being a former French colony. The first exporting company, *Société Beninoise de Pêche*, was established in 1993 and its first exports to the EU followed soon. The export connection was made official with the inscription of Benin for the import of fishery products on June 30, 1998 on part II of the list annexed to the Decision 97/296/EC, after a documentary assessment of the application for approval was sent to the European Commission. About 150 shrimp collectors were trained to buy raw shrimps for the processing plants. They selected the shrimps that fulfilled the export requirements, while the rejected shrimps were sold on domestic markets. The raw shrimps were transported to one of the three processing plants located in

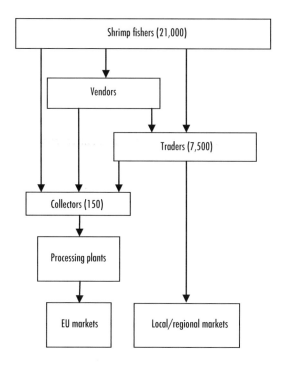

Figure 10.1. The Beninese shrimp chain (2012).

Cotonou, the economic capital of Benin, where they were processed and exported to the EU as raw frozen shrimps.

Until 2001, Benin's shrimp exports grew rather steadily, increasing from 333 tons in 1995 to 734 tons in 2001. After a visit of experts from the European Food and Veterinary Office (FVO) in 2002, the situation would, however, change dramatically. The delegation came to inspect the Beninese export chain for shrimps in order to advise the EU on the potential hazards associated with Beninese shrimps for European consumers. The mission evaluated the 'equivalence of the legislation of Benin with corresponding Community requirements and performance of national authorities for the control of conditions of production and export of fishery products to the EU' (OAV, 2002). The inspection involved the whole production chain, following the farm-to-fork approach. The inspectors visited several fishermen, the fishing ports, the landing sites, the three shrimp processing plants, the National Health Laboratory as well as the Competent Authority (CA). In the end, the mission pinned down deficiencies in the conditions of production themselves[1], as well as the lack or obsolescence of legal measures[2] supposed to be the basis for a control system aiming at the production of safe end products. As a result, Benin was declared having no legal basis to guarantee and certify the compliance of shrimps exported to the EU with the requirements of Directive 91/493/EEC then in force within the EU.

In reaction to this inspection, Beninese authorities immediately provided the FVO with an action plan with a detailed timetable and submitted an application for funding to the European Commission. However, these guarantees were deemed insufficient, leading Beninese authorities to decide to self-ban shrimp exports to the EU with the aim to undertake remedial measures. The self-ban was maintained for eighteen months up till February 2005.

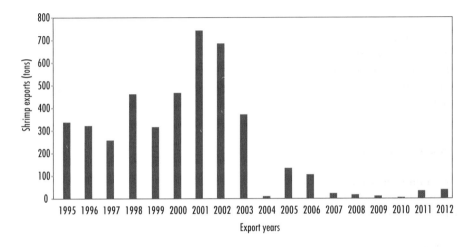

Figure 10.2. Shrimp exports (in tons). Compiled from the Directorate of Fisheries' database (2013).

---

[1] These refer overall to hygiene and lack of knowledge of good production practices, including the implementation of HACCP principles in shrimp processing plants. HACCP is a systematic and science-based approach that enables the detection of potential food safety hazards in order to take preventive measures. The adoption of a HACCP procedure consists of firstly, identifying the Control Critical Points (CCPs) in preparation, processing and distribution stages as the basis for the control of potential hazards in foodstuffs and secondly, establishing a system for their prevention, elimination or reduction to acceptable levels.

[2] The 6 texts then presented were still in draft form and contained gaps in the definition of essential parameters such as maximum limits for contaminants in the natural environment, histamine and additives, the list of food contact materials, cleaning and desinfection products allowed. Consequently, there was a lack of control procedures with regard to the organisation of landing activities, the collection of freshly caught shrimps, the uneven implementation of production principles, and the use of questionable methods, including hyperchlorination, by operators.

After the self-ban, the exports decreased from 680 tons in 2002 to 96 tons in 2006 and 2 tons in 2009 (Figure 10.2). In that year, a second visit of the FVO to Benin was carried out to assess the reforms and to decide whether the system in Benin had become equivalent to that of the EU. Although this mission acknowledged the significant improvements in the sector since 2002, both in institutional and legislative terms, only the legislation was declared equivalent to EU requirements, while the control system and the functioning of the processing plants were not deemed equivalent to relevant parts of EU law. Despite these non-compliances, the decision 2009/951/EU confirmed Benin as a country from which imports of fishery products for human consumption within the EU is authorised as from January 2010 (EC, 2009).

The burden on the three shrimp processing plants that exported before the ban had, however, been too heavy and they were forced to stop their operations. During the 2012 fishing season, only one processing plant was operational. This plant had been newly established after the ban and, interestingly, it does not just export to the EU, but also to China. With only one factory being operational, shrimp fishers are largely constrained to domestic markets. Based on data collected from 183 shrimp fishers, Adekambi et al. (2012) found that about 71% of the total shrimp production in 2012 was sold within Benin. Only 17% of this 2012 production was exported (the remaining 12% was consumed by the fishers, or was wasted).

10.5 Factors affecting microbial contamination

Environmental factors and shrimp handling practices have an important impact on the microbiological quality and safety of Beninese shrimps. Both factors affect the microbial contamination level of freshly caught shrimps and their shelf life.

10.5.1 Environmental factors

Shrimp fishing areas are located in the coastal region of the country characterised by a high population density due to the presence of the most important cities of Benin, such as Cotonou (the economic capital), Porto-Novo (the administrative capital), and the suburb Abomey-Calavi. Waste management in these cities remains problematic. The banks of the channel of Cotonou, which connects the Lake Nokoue to the Atlantic Ocean, are occupied by numerous dumps of garbage and discharge points of industrial waste water (Adeke et al., 2011); Badahoui et al., 2009). The huge Dantokpa fresh produce market near Lake Nokoue is also a source of pollution. The waste produced at this market, which is the trading hub of Western and Central African countries (Dossou and Glehouenou-Dossou, 2007), are drained into the lake through rainwater. Moreover, the population of lacustrine cities (cities established in the lake) in Lake Nokoue without a proper management of solid and liquid waste is increasing. For instance, the high total organic carbon recorded in sediment samples from Ganvie, a lacustrine city located in Lake Nokoue, could be attributed to the fact that all the waste that the inhabitants of this city generate are continuously thrown into the lake without treatment (Soclo *et al.*, 2002). The lakes and lagoons are thus waste receivers and, therefore, affect the microbial contamination level of shrimps. Obviously, the microbial status of seafood after catch is closely related to environmental conditions and microbiological quality of the water (Feldhusen, 2000). For instance, the high initial microbial contamination load of shrimps from Lake Nokoue and Lake Aheme is mainly due to the water quality of these lakes (Degnon *et al.*, 2012; Dossou *et al.*, 2007).

10.5.2 Shrimp handling practices

Our investigations in 2011 revealed that shrimp fishing generally takes place at night or early in the morning (4:00-6:00 AM). Among the used fishing nets, the drift net remains the most popular. Using this fishing net, three or four successive captures are realised and poured into the same container. About 4 hours elapse between the first catch and the last one. The most frequently used storage containers by shrimp fishers are their wooden boats. Shrimps are stored in these boats at ambient temperature (±27 °C) before delivery to intermediate traders. Most of the intermediate traders just wash the shrimps using lake or lagoon water, and also keep their products at ambient temperature in baskets during approximately 1.5 hours before delivery to the collectors who finally store them on ice. The collected shrimps are then transported to Cotonou where the shrimp processing plants are located.

There are various shortcomings related to post-catch shrimp handling prior to processing. The practice of dumping caught shrimps in the wooden boats where different catches are mixed does not meet the requirements of European Regulation (European Commission, 2004a), which states that each successive catch must be kept separately. Moreover, storing shrimps at ambient temperature from catch until delivery to collectors is in violation of the requirements of the Codex Alimentarius (CAC, 2003) and European Regulations (European Commission, 2004a) which stipulates that chilling of fishery products should start as soon as possible. In fact, shrimps are extremely sensitive to deterioration (Mendes *et al.*, 2001). Unlike other crustaceans (crabs, lobsters) which can be kept alive until processing, shrimps die soon after capture, and they are often contaminated with bacteria from their endogenous microflora as well as from the mud trawled up with them (Adams and Moss, 2006). Therefore, the storage temperature is crucial (Cyprian *et al.*, 2008; Matches, 1982). For instance, sensory analysis done on fresh samples showed that raw chilled shrimps reached the limit of acceptability (50% rejection) after four days in ice, whereas 100% of the samples stored at room temperature (22 °C) were rejected after 24 hours (Mendes *et al.*, 2001). In addition, the quality of lake/lagoon water used to wash shrimps is not in agreement with European regulation since non-potable water is a source of contamination (European Commission, 2004a).

Also the conditions in which shrimps were processed at processing plants were below standards. The FVO experts revealed several shortcomings including a lack of testing the effectiveness of cleaning and disinfection procedures, a lack of water pipes to drain used water, the presence of rusty materials, and non-compliance with the HACCP plan (Office Alimentaire et Vétérinaire, OAV, 2002).

10.6 Factors affecting chemical contamination

Shrimps are susceptible to chemical contamination, in particular through environmental pollution and post-catch handling.

10.6.1 Environmental factors

The excessive use of chemical fertilisers and pesticides for cotton production upstream from the fishing areas leads to the pollution of the lakes and lagoons. In fact, the chemical products are drained by rainwater into the rivers, which in turn flow into the lakes and the lagoons (Adeke *et al.*, 2011). For instance, chemical analyses on the flesh of tilapia from the Oueme river which springs out of Taneka hills in the Northen part of Benin and flows into Lake Nokoue, showed that different organochlorine pesticides (including banned pesticides) were present in their tissues (Okoumassoun *et al.*, 2002). The average concentration of cadmium and lead in fish samples, other than shrimps, collected from Lake Nokoue was higher than the allowed limit (Hounkpatin *et al.*, 2012). Kaki *et al.* (2011) found that both in water and sediment from Lake Nokoue, concentrations of some toxic metals such as cadmium, lead, copper, and arsenic were high and sometimes exceeded the standards. With regard to shrimps specifically, research showed that unlike other metals, such as cadmium and zinc, the concentration of lead in shrimps from Lake Nokoue exceeded the standard values (Aina *et al.*, 2012). The analyses of heavy metals carried out in 2007 on shrimps from Lake Aheme revealed that the concentrations of lead, cadmium and mercury were all lower than the allowed limits. However, the concentration of mercury was very close to the permissible limit (Coopéartion Belge au dévelopement, 2007). The lower level of metals in Lake Aheme could be explained by the remote geographical location of this lake, at large distance from the big cities.

The discharge of petroleum-based products in the Lake Nokoue (Dovonou *et al.*, 2011) is another important source of chemical contamination since these products contain polycyclic aromatic hydrocarbons (PAH). Also, the common practice of washing cars at the lake shores contributes to their pollution (Adeke *et al.*, 2011).

10.6.2 Shrimp processing

In order to preserve the freshness of their products, processors use chemical products, such as chlorine and sulphites. During their first visit in 2002, the FVO experts pointed out that it is not allowed to use chlorine to decontaminate raw shrimps in processing plants. They also reported the improper use of sulphite by shrimp processors (OAV, 2002). Even in 2009, the FVO experts recommended that the processing plants should better control the use of additives (OAV, 2009).

10.7 The institutional environment

Some of the factors that lead to the microbial and chemical contaminations are rooted in the institutional environment surrounding the chain. Chain actors face a lack of access to productive resources, lack of institutional support, and legal constraints. Productive resources include, among others, capital, knowledge and skills. Actors in the sector, in particular shrimp fishers, face serious capital constraints which limits their capacity to invest. For instance, in 2012 only 6% of shrimp fishers that we interviewed had access to credit during the previous five years. Furthermore, the literacy level among shrimp fishers is low. About 49% of the shrimp fishers did not receive formal education. For those shrimp fishers that are educated, the average number of years at school is 5, which is equivalent to primary level.

Institutional support includes provision of market information and basic infrastructure, like roads that are suitable for transporting shrimps. The shrimp chain actors in Benin lack information on the preferences of EU consumers. The little information that is available is often not shared between actors, because relationships are weak. This leads to market information asymmetry preventing the chain from responding to market demands. Unlike traders, shrimp fishers hardly know about changes in the market value of their product. Furthermore, only 29% of shrimp fishers interviewed in 2012 reported receiving advice from extension services (Adekambi *et al.*, 2012). Shrimp fishers who have no contact with extension services are lacking market information, as extension services represent one of the main sources of market information for small-scale farmers (e.g. Marter, 2005). Being isolated from basic market information, shrimp fishers also have few incentives to take measures that may improve the quality of the shrimps.

Lack of access to proper infrastructure, such as connections with main roads and markets, constitute another impediment to the integration with export markets. About 63% of shrimp fishers interviewed in 2012 lived at least at 5 km from the nearest main market, and 35% of those lived at least at 10 km from the market. Likewise, 92% of the surveyed shrimp villages were about 4 km from main roads (Adekambi *et al.*, 2012). Despite this proximity to main roads, the accessibility of villages still remains an issue. For instance, 32% of the shrimp villages surveyed by Adekambi *et al.* (2012) are entirely surrounded by water and 45% of them are difficult to access during the rainy season. Such difficulties in accessing main roads hamper the market information flow to shrimp fishers and prevent them from engaging in more distant markets because of high transaction costs.

Legal barriers for Benin producers derive substantially from the normative requirements imposed by the EU, the main export market. The current EU food regulatory regime was established in 2002 by the regulation EC No 178/2002, commonly called General Food Law (GFL) that entered into force in January 2004. The GFL is a comprehensive and integrated approach to food safety, which covers any stage of production, processing and distribution of food, following a farm-to-fork approach. In 2004, a set of complementary regulations was adopted, commonly referred to as the 'food hygiene package'[3]. EU food law combines requirements derived from transparency and proportionality with, wherever applicable, measures based on the principles of risk analysis, which include risk assessment, risk management and risk communication. When following an assessment of available information, the possibility of harmful effects on health is identified but scientific uncertainty persists, provisional measures must be adopted as part of risk management. The GFL establishes the primary responsibility of Food Business Operators (FBO's) to ensure that food placed on the market meets the requirements of food law, therefore entrusting them for the implementation of HACCP principles, including the obligations of traceability and own checks thereon.

With regard to third countries, 'Food and feed imported into the Community for placing on the market within the Community shall comply with the relevant requirements of food law or conditions recognised by the Community to be at least equivalent thereto or, where a specific agreement exists between the Community and the exporting country, with requirements contained therein' (European Commission, 2002). For products of animal origin, the most important import standard is the formal approval requirement of the third country or a part of it. In that respect a list of authorised countries is drawn up and updated by the Community according to a wide range of criteria generally comprising the ability of the legislation to comply with general and specific Community requirements, the financial and material capacity of so-called Competent Authorities to implement, enforce and properly control rules set up by that legislation (European Commission, 2004c). The detection of excess of contaminants and residues, the non-compliance with relevant legislation on organoleptic, chemical, physical or microbiological elements, the presence of poisons are all sufficient grounds to declare a fishery product unfit for human consumption. It is therefore in view of these principles that the EU regularly delegates missions to assess the compliance of exporters with its standards. To be granted access to the EU, the Competent Authority of the exporting country should control and inspect all stages of the chain.

---

[3] Food hygiene refers to 'the measures and conditions necessary to control hazards and to ensure fitness for human consumption of a foodstuff taking into account its intended use'. (Reg. 852/2004, art. 2.a). The current hygiene legislation comprises the Regulations 852/2004 (setting basic rules and principles applicable to food business operators throughout the food chain), 853/2004 (defining specific hygiene rules to be applied to food of animal origin), and 854/2004 (setting the rules to be applied by national authorities for the control and enforcement of food hygiene policies), along with Regulation 882/2004 (defining an EU-wide harmonised approach to the design and implementation of national control systems).

10.8 Measures for improvement

10.8.1 Legislative measures

Originally, Law 84-009 on the control of foodstuffs, adopted in 1984, was applicable to shrimp consumed in Benin and exported from Benin. However, due to the lack of implementing decrees, this law was never properly applied. Confronted with the ambition to export to the EU and after the shortcomings indicated by the FVO, at the beginning of this century, Benin was compelled to adopt and implement a more effective food law. Considering the needs of the moment, regulatory steps, instead of defining a broad framework for food products in general, were rather focused on fishery products. The radical changes that ensued cover both the legal and institutional environments, even though doubts still persist about their effectiveness.

Triggered by the FVO mission, legal reforms in Benin evolved following a process in two phases. The first phase built upon Law 84-009 on the control of foodstuffs. It started by the adoption in 2003 of the Decree 114 on the quality assurance of fishery products to implement the Law 84-009. Quality assurance is defined as all the concerted and systematic measures necessary to obtain reasonable assurance that a product or service meets given quality requirements. In that respect, this decree defines the regime of the health rules for fishery products, their conditions of processing, preservation, and marketing, as well as the tax system related to these products. It paved the way for the adoption of a set of seven other orders in the same year that deal with questions related to processing plants, hygiene, good laboratory practices and to limits of certain substances in fishery products. The violations by establishments of relevant provisions are sanctioned by a withdrawal or suspension of approval (République du Bénin, 2003b). It is worth noting that these texts were aiming at complying with Directive 91-493 (European Commission, 1991) which was in force at the time of the first FVO inspection in 2002.

Following the change in EU law in 2004 (more particularly the adoption of the GFL), the second phase of legal reforms in Benin departed from the former pattern and covered a broader set of laws applicable to all foodstuffs in general. This process resulted in 2009 in the adoption of five new texts, which literally copy EU current legislation, therefore subscribing to all its requirements. This step was made necessary due to the holistic approach introduced in EU law through the GFL. It proved essential to integrate law concepts, such as risk analysis or traceability in Benin, and better define responsibilities with regard to food safety.

10.8.2 Control system measures

Without a well-organised and properly functioning Competent Authority, there could be no possibility of granting a country or its producers the authorisation to export food to the EU. In Benin, the most important institution remains the Directorate of Fisheries, acting as the Competent Authority according to the appellation required by the EU. In this respect, the Competent Authority must ensure the registration of shrimp fishers, the implementation of hygiene rules at landing sites, the proper functioning of processing plants, and organise the necessary check of end products so as to deliver the health certificates that authorise the export to the EU.

A new system of shrimp collection is also being implemented, termed Improved System of Shrimp Collection for Export (ISSCE). Two types of landing sites have been constructed, a so-called Plat-Form of Transfer (PFT) and the Bases of Compulsory Check (BCC). The latter are larger and better equipped than the first. According to this new system, shrimp fishers that have cooling boxes are supplied with ice at PFT level before going out to fish. The captured shrimps are immediately stored on ice in the cooling boxes. After fishing, shrimp fishers return to PFT to sell their catch to intermediate traders or collectors. At this level, shrimps are washed, sorted and put on ice in standardised isothermal boxes. The packed products are forwarded to the BCC where a qualified inspector checks them and delivers a basic sanitary certificate. The boxes are then transported to the processing plants where they are cleaned after the unloading of the shrimps. The cleaned boxes return to the BCC where they are filled with ice and sent to the PFT where the Standardised Isothermal Boxes (SIB) are filled with the catch bought by the collectors or intermediate traders and the cycle resumes. So far, two BCC and 12 PFT have been built at the shores of Lake Aheme. Only 3 PFT and 2 PFT are under construction at Lake Nokoue and the lagoon of Porto-Novo, respectively.

10.8.3 Measures at processing plants

Processing plants play a crucial role in the system, considering that ultimately the whole production process remains under their responsibility. They must carry out their own organoleptic checks while receiving supply of fresh shrimps from collectors, put in place a functioning traceability system, and implement HACCP principles.

The self-ban on exports was not without consequences for the companies. For one and a half years they were cut off from their main market, while at the same time investments (more particularly in order to comply with EU requirements) were demanded. In 2005 the EU again allowed imports of Beninese shrimps, but not after it demanded one more sacrifice from the processing plants, namely to destroy all their stocks as precautionary measure. As a consequence, 189 tons of shrimps valued at 700 million CFA francs (about 1.1 million Euros), were destroyed in early 2005. While 175 tons were nevertheless exported in 2005, exports fell to 32 tons in 2006. The following years, however, were not successful, with respectively 54.6 and 6.2 tons exported in 2007 and 2008. The main cause of the drop of exports seems the financial position of the companies that had been seriously affected by the self-ban. In 2009, one processing plant had closed down, the premises of the second were seized by the tax administration, and the third one was closed for renovation. In 2012, all processing plants that were active before the self-ban closed down and only one, newly established, processing plant was exporting shrimps. Interestingly, this new processing plant not only targets the EU, but also the Chinese market.

The state of the processing plants also affected control practices. Although during their last visit in June 2009, the FVO experts noted considerable progress, they also raised shortcomings related to the adoption and application of HACCP plans (OAV, 2009). In addition, four notifications (three in 2005 and one in 2006, all from Spain) from the Rapid Alert System for Food and Feed of the EU, mentioned a too high content of sulphite in frozen whole raw shrimps (*Penaeus* spp.) (RASFF, 2011). However, to date, no analyses have been carried out on exported shrimps to evaluate the health risks associated with their consumption (Le Ry, 2007).

10.8.4 Responsiveness of the shrimp fishers to the measures

In the end, the effectiveness of the processing plants depends not only on their own practices, but to a large extent also on the practices of the shrimp fishers whose behaviour substantially influences the quality and safety of the shrimps.

A study by Adekambi *et al.* (2012) gave a deeper insight in the patterns of responsiveness of shrimp fishers, as well as on their drivers and consequences. This study measured reactive and proactive attitudes of fishers towards the market. It found that shrimp fishers' (reactive) responsiveness was 2.13 on a 5-point scale, indicating that shrimp fishers on average respond modestly to processing plants' demands (Adekambi *et al.*, 2012).

The results also show that the reactive behaviour is embedded in strong social relationships with traders. In most developing countries, regulative institutions, such as governments and legislation are less effective in managing illicit aspects of the marketing system (e.g. Burgess and Steenkamp, 2006). Shrimp fishers and traders generally rely on each other's trustworthiness to run and sustain their businesses. Traders generally share market information with the shrimp fishers they trust. Furthermore, traders actively train these trusted shrimp fishers in specific fishing, treatment and handling skills. They also provide trusted shrimp fishers with loans at the beginning of each fishing campaign and offer fixed purchasing prices to their shrimp providers. Such relationships enable shrimp fishers to take effective fishing decisions, such as adopting cooling boxes and ice to meet their customers' expectations. Fishers who do not have such trust-based relationships with traders or whose ongoing relationships do not provide such support take a more proactive attitude in order to engage in a partnership with a more trustworthy trader.

The behaviour of both reactively and proactively market-oriented shrimp fishers is supported by the norms and values of the communities in which they live. Such support is important, because shrimp fishers live at or near subsistence levels and depend on social relationships for their investments in fishing gear and extra labour. In particular, the communities that are better connected to the road system have developed such supportive cultures. As their physical connection to the export market is easier to establish, they have probably benefitted more from the exports in the past and thus developed a culture that supports fishers that aim to comply with the standards of exporting companies. For instance, peers who keep telling shrimp fishers to understand and satisfy customers' wants, are associated with a high level of shrimp fishers' responsiveness to customers' expectations.

Notably, export market integration still pays off for the fishermen. The shrimp fishers that sell a larger share of their shrimp catch to the export-oriented marketing channel, exhibit higher levels of livelihood in that they report to be wealthier and happier than others (Adekambi *et al.* 2012). Shrimp fishers that export their shrimps receive higher prices than those that sell to domestic marketing channels, because exporting companies are willing to pay premium prices for good quality shrimps. Such higher prices lead to higher income that, in turn, contributes to improved livelihoods (see also Maertens and Swinnen, 2009).

10.9 Discussion

Looking back, the Beninese export chain for shrimps made a false start because of a mismatch between the safety levels required by the EU and the level that the Beninese chain members could jointly deliver.

The inconsistency between the provisions of the laws and their implementing decrees has been indicated as the main weakness of the food safety law in Benin (Amaskame, 2010). These accusations, however, neglect that the step from meeting the requirements for a local market and those of the EU market, may have been too big to make at once. On hindsight, the food safety and quality problems should probably have been anticipated when the chain was created in the 1990s, because repairing the system afterwards appeared a painful process in which much of the prior investments were destroyed.

Undoubtedly, the protection of consumers is a legitimate objective, and such a vision must be shared by all countries regardless of their level of development. However, as far as developing countries are concerned, an approach that better balances their needs and priorities should be advocated. A strategy by which the chain could improve step by step, developing from local, via regional (other African), to high-income export markets like the EU might have been a more sustainable approach to overcoming the constraints arising from EU requirements. At the very least, this would not have led to the near disappearance of the shrimp export sector. The recent start of exports to China is in that respect probably a wise decision as it makes the sector less dependent on a single export market.

While most reforms were supported by the financial and technical assistance of the EU, they would have certainly be more beneficial with complementary economic and marketing expertise that would better frame their implementation. Looking back, the desperate attempts to comply with EU requirements to re-establish market access have probably done more harm than good. Ironically, with a substantial part of the chain being destroyed, the EU market has opened up again. Whether the access is sustainable is another question. Several measures will be necessary to prevent that history will repeat itself.

10.10 Recommendations

It has been widely acknowledged that standards are increasingly acting as important assets to enhance economic welfare. International standards may lead to greater efforts at capacity building for developing countries (Charnovitz, 2002), thereby modernising export-oriented chains, and improving health standards at national level, with positive spill-over effects on competitiveness and market access (Jaffee and Henson, 2004). It emerges that only an actual improvement of production and inspection conditions could guarantee and sustain international market access. In the case of Benin, some steps are still necessary to ensure the efficiency of the shrimp export sector.

To maintain access to the European market, it is first of all important to better manage the solid and liquid wastes around the lakes and the lagoon. The Competent Authority should pursue the implementation of monitoring plans for aquatic contaminants. Shrimp handling conditions should also be improved. In this respect, the recently implemented system of shrimp collection (ISSCE) will be useful.

Second, investments will be necessary both in the private and public domain. It is recommended to pursue investment in infrastructure in order to situate the landing sites as close as possible to the fishing areas to improve shrimp handling practices. Laboratories must exist with the relevant capacities to carry out chemical and microbiological tests to certify compliance with safety rules. A further step would be to work towards accreditation of those laboratories to gain reliability on the international market. Work in progress such as the census of shrimp fishers must be fully completed to facilitate the implementation of the traceability requirements.

In the private domain, shrimp fishers themselves need to develop the appropriate competencies, i.e. knowledge and skills, not only to generate and exploit knowledge about customers' needs and wants, but also to deploy their scarce resources in more competent ways. Besides, to further boost such competencies, traders should reconsider their relationships with shrimp fishers. The development and maintenance of strong relationships, may be a productive strategy in enhancing shrimp fishers' integration with export chains. Moreover, traders could equally enforce quality standards upon the fishers, thus providing better guarantees of compliance with the stringent requirement in the EU or other targeted export markets. Finally, activities that encourage communities' emphasis on responding to customers' needs and wants must be supported by policy makers and/or NGOs and development projects.

Third, the training of actors in Good Hygiene Practice, Good Fisheries Practices, and audits on implementation of HACCP procedures, should be regularly organised by the authorities in charge of the sector. This is of importance for shrimp processors. In addition to enhance food safety and thus confidence in food trade, the benefits of applying the HACCP system include better use of resources and more timely response to production systems (Moy *et al.*, 1994).

Fourth, research could also play an important role in improving shrimp quality. Little is known about the specific hazards associated with shrimps that could threaten consumer's health. To our best knowledge, the main pathogenic bacteria associated with Beninese shrimps are yet unknown. In addition, the real threats of heavy metals and pesticides related to the consumption of Beninese' shrimps have not yet been elucidated. In fact, the results of the technical inspection carried out by the FVO experts prompted the national government in 2003 to self-ban the shrimp export as a precautionary action. To date, no analyses have been carried out on exported shrimps to evaluate the health risks associated with their consumption (Le Ry *et al.*, 2007). Also a mathematical model describing the effects of temperature and time in the chain on the concentration of relevant microorganisms associated with shrimps could be developed. Such a model will facilitate advising stakeholders on the most effective measures to be taken in the chain to improve safety and quality of shrimps. Furthermore, from both resource-based and marketing perspectives, there is a need to reconsider shrimp fishers' relationship to market integration by investigating competence values that might enable them to develop capabilities and accumulate resources necessary to identify and satisfy customers' needs and wants.

Fifth, aquaculture could be a good option for Benin. This will enable not only a better control of the quality of its products, but also ensure the continuous operation of shrimp processing plants. In fact, the current seasonal character of the fishing of wild shrimps does not make it possible for processing plants to ensure the full use of their processing lines.

Lastly, it might be interesting to raise more awareness about the safety of food at the national level. This will generate from the inside the demand for the law, and bring stakeholders together towards the same goal of a better implementation and enforcement of the food law. Only in such a way could the past lethargy be avoided, and the EU and other export markets be 'captured' sustainably.

Acknowledgments

This case was developed in the context of NPT projects 262 and 263. The authors thank NUFFIC for their financial support. They are also grateful to Jos Bijman and Verena Bitzer for their valuable comments and suggestions.

References

Ababouch, L., 2006. Assuring fish safety and quality in international fish trade. Marine Pollution Bulletin 53: 561-568.

Ababouch, L., Gandini, G. and Ryder, J., 2005. Causes of detentions and rejections in international fish trade. FAO Fisheries Technical Paper 473. FAO, Rome, Italy.

Adams, M.R. and Moss, M.O., 2006. Food microbiology. Royal Society of Chemistry, London, UK.

Adekambi, S.A., Ingenbleek, P.T.M. and Van Trijp, H.C.M., 2012. Integrating subsistence producers from developing and emerging countries with the world market: do reactive and proactive market orientation matter? Working paper, Wageningen University, Wageningen, the Netherlands.

Adeke,T.B., Houssa, R., Koukou, D.H., Tossou, S. and Verpoorten, M., 2011. Stratégies de relance de la filière crevette au Bénin. Direction des peches, Benin.

Aina, M.P., Degila, H., Chikou, A., Adjahatode, F. and Matejka, G., 2012. Risk of intoxication by heavy metals (Pb, Cd, Cu, Hg) connected to the consumption of some halieutic species in lake Nokoue: case of the *Penaeus* shrimps species and the *sarotherodon melanothereon*. British Journal of Science 5: 104-118.

Amaskane, M., 2010. Appui aux systèmes d'information sur les mesures sanitaires et phytosanitaires au Bénin. Rapport de Mission MTF/BEN/053/STF, 73 pp.

Aregheore, E.M., 2009. Country pasture/forage resource profiles: the republic of Benin. FAO, Rome, Italy. Available at: http://tinyurl.com/pw3axtm.

Badahoui, A., Fiogbe, E. and Boko, M., 2009. Les causes de la dégradation du chenal de Cotonou. International Journal of Biological and Chemical Sciences 3: 979-997.

Baffes, J., 2011. Cotton subsidies, the WTO, and the 'cotton problem'. World Economy 34: 1534-1556.

Bernard, T., Taffesse, A.S. and Gabre Madhin, E., 2008. Impact of cooperatives on smallholders' commercialization behavior: evidence from Ethiopia. Agricultural Economics 39: 147-161.

Birthal, P.S., Joshi, P.K., Roy, D. and Thorat, A., 2007. Diversification in Indian agriculture towards high-value crops. The role of smallholders. Markets, Trade and Institutions Division. Discussion Paper 00727. International Food Policy Research Institute (IFPRI), WAshington, DC, USA.

Bougheas, S., Demetriades, P. and Morgenroth, E., 1999. Infrastructure, transport costs and trade. Journal of International Economics 47: 169-189.

Burgess, S.M. and Steenkamp, J.-B.E.M., 2006. Marketing renaissance: how research in emerging markets advances marketing science and practice. International Journal of Research in Marketing 23: 337-356.

Buys, P., Deichmann, U. and Wheeler, D., 2010. Road network upgrading and overland trade expansion in sub-Saharan Africa. Journal of African Economies 19: 399-342.

Castells, M. and Portes, A., 1989. World underneath: the origins, dynamics, and effects of the informal economy. In: Portes, A., Castells, M. and L.A. Benton (eds.) The informal economy: studies in advanced and less advanced developed countries. Johns Hopkins University Press, Baltimore, MD, USA, pp. 11-37.

CGIAR, 2005. CGIAR system research priorities 2005-2015. CGIAR Science Council Secretariat. Available at: http://tinyurl.com/z9jv6bn.

Charnovitz, S., 2002. International standards and the WTO. The George Washington University Law School Public Law and Legal Theory. Working Paper, 32 pp.

Codex Alimentarius Commission, 2003. Code of practice for fish and fishery products. CAC/RCP 52: 1-134. Available at: http://tinyurl.com/p7q7v72.

Coopération belge au dévelopement, 2007. Dossier technique et financier: appui au dévelopement des filières halieutiques du Bénin. BEN 06 013 11. Available at: http://tinyurl.com/zvb2yuo.

Cyprian, O.O., Seveinsdòttir, K., Magnùsson, H. and Martinsdòttir, E., 2008. Application of quality index method (QIM) scheme and effects of short-time temperature abuse in shelf-life study of freshwater arctic char (*Salvelinus alpinus*). Journal of Aquatic Food Product Technology 17: 303-321.

Diaz Rios, L.B. and Jaffee, S., 2008. Barrier, catalyst, or distraction? Standards, competitiveness, and Africa's groundnut exports to Europe. Agriculture and Rural Development Discussion Paper, World Bank, Washington, DC, USA.

Degnon, R.G., Dahouenon-Ahoussi, E., Adjou, E.S., Ayikpe, O., Tossou, S., Soumanou, M.M. and Sohounhloue, D.C.K., 2012. Impact des traitements post-capture sur la qualité microbiologique des crevettes (*Penaeus* spp.) du lac Ahémé au Bénin destiné à l'exportation. Journal of Applied Biosciences 53: 3749-3759.

Délégation de la Commission Européenne au Bénin, 2009. Assistance technique pour une etude diagnostique sur le développement du secteur privé au Bénin et l' identification d' un projet de compétitivité et croissance sous le 10eme FED/Diagnostic de la filiere crevette, Aria Consult, 50 pp.

Deshingkar, P., Kulkarni, U., Rao, L. and Rao, S., 2003. Changing food systems in India: resource sharing and marketing arrangements for vegetable production in Andhra Pradesh. Development Policy Review 21: 627-639.

Dossou, J., Tobada, P., Sedogbo, Y.A., Mama, D., Tossou, S., Ouikoun, G., Laleye, P. and Capochichi, B., 2007. Impact de la pollution de l'environnement sur la qualité sanitaire des crevettes capturées sur les pêcheries du lac Nokoué. Annale de la Faculté des Sciences Agronomiques du Bénin, Université d'Abomey-Calavi 5: 123-127.

Dossou, K.M.R. and Glehouenou-Dossou, B., 2007. The vulnerability to climate change of Cotonou (Benin): the rise in sea level. International Institute for Environment and Development 19: 65-79.

Dovonou, F., Aina, M., Boukari, M. and Alassane, A., 2011. Pollution physico-chimique et bactériologique d'un écosystème aquatique et ses risques écotoxicologiques: cas du lac Nokoué au Sud Benin. International Journal of Biological and Chemical Sciences 5: 1590-1602.

European Commission, 1991. Council Directive 91/493/EEC laying down the health conditions for the production and placing on the market of fishery products. Official Journal of the European Union L 268: 15-34.

European Commission, 1997. Commission Decision 97/296/EC drawing up the list of third countries from which the import of fishery products is authorized for human consumption, Official Journal of the European Union L 122: 21-23.

European Commission, 2002. Regulation 178/2002 laying down the general principles and requirements of food law, establishing the European Food Safety Authority and laying down procedures in matters of food safety. Official Journal of the European Union L 31: 1-24.

European Commission, 2004a. Regulation 852/2004 on the hygiene of foodstuffs. Official Journal of the European Union L 139: 1-54.

European Commission, 2004b. Regulation 853/2004 laying down specific hygiene rules for food of animal origin. Official Journal of the European Union L 139: 55-205.

European Commission, 2004c. Regulation 854/2004 laying down specific rules for the organization of official controls on products of animal origin intended for human consumption. Official Journal of the European Union L 139: 206-320.

European Commission, 2004d. Regulation 882/2004 on official controls performed to ensure the verification of compliance with feed and food law, animal health and animal welfare rules, Official Journal of the European Union L 165: 1-141.

European Commission, 2005. Regulation 2076/2005/EC laying down transitional arrangements for the implementation of Regulations (EC) No 853/2004, (EC) No 854/2004 and (EC) No 882/2004 of the European Parliament and of the Council and amending Regulations (EC) No 853/2004 and (EC) No 854/2004, OJ L338, 83-88, amended by Regulation 1666/2006. Official Journal of the European Union L 320: 47-49.

European Commission, 2009. Commission decision 2009/951/ EU amending annexes I and II to Decision 2006/766/EC establishing the lists of third countries and territories from which imports of bivalve molluscs, echinoderms, tunicates, marine gastropods and fishery products are permitted. Official Journal of the European Union L 328: 70-75.

Food and Agriculture Organization of the United Nations (FAO), 2003. FAO's strategy for a food chain approach to food safety and quality: a framework document for the development of future strategic direction. Committee on Agriculture, Seventeenth Session, March 31-April 4, 2003, COAG/2003/5. Available at: http://tinyurl.com/jax7qda.

Feldhusen, F., 2000. The role of seafood in bacterial foodborne diseases. Microbes and Infection 2: 1651-1660.

Gulati, A, Minot, N., Delgado, C. and Bora, S., 2007. Growth in high-value agriculture in Asia and the emergence of vertical links with global dupply chains, standards and the poor: how the globalization of food systems and standards affects rural development and poverty. CABI, Wallingford, UK, pp. 91-108.

Henson, S. and Loader, S., 2001. Barriers to agricultural exports from developing countries: the role of sanitary and phytosanitary requirements. World Development 29: 85-102.

Hounkpatin, A.S.Y., Edorh, A.P., Salifou, S., Gnandi, K., Koumolou, L., Agbandji, L., Aissi, K.A., Gouissi, M. and Boko, M., 2012. Assessment of exposure risk to lead and cadmium via fish consumption in the lacustrian village of Ganvie in Benin Republic. Journal of Environmental Chemistry and Ecotoxicology 4: 1-10.

Jaffee, S. and Henson, S., 2004. Standards and agro-food exports from developing countries: rebalancing the debate, World Bank policy research. Working paper 3348. World Bank, Washington, DC, USA.

Jongwanich, J., 2009. The impact of food safety standards on processed food exports from developing countries. Food Policy 34: 447-457.

Kaki, C., Guedenon, P., Kelome, N., Edorh, P.A. and Adechina, R., 2011. Evaluation of heavy metals pollution of Nokoue lake. African Journal of Environmental Science and Technology 5: 255-261.

Kambewa, E.V., 2007. Balancing the people, profit and planet dimensions in international marketing channels. a study on coordinating mechanisms in the Nile perch channel from Lake Victoria. PhD Thesis, Wageningen University, Wageningen, the Netherlands.

Le Ry, J., Barry, O. and Legendre, E., 2007. Plan de relance de la filière halieutique. Projet d'Appui au Secteur Privé (PASP). Programme: BEN/009/004. Available at: http://tinyurl.com/zbzdubg.

Licht, A.N., Goldschmidt, C. and Schwartz, S.H., 2005. Culture, law, and corporate governance. International Review of Law and Economics 25: 229-255.

Maertens, M., Colen, L. and Swinnen, J.F.M., 2011. Globalisation and poverty in Senegal: a worst case scenario? European Review of Agricultural Economics 38: 31-54.

Maertens, M. and Swinnen, J.F.M., 2009. Trade, standards, and poverty: evidence from Senegal. World Development 37: 161-178.

Marter, A., 2005. Market information needs of rural producers. In: Almond, F.R. and Hainsworth, S.D. (eds.) Beyond agriculture-making markets work for the poor: proceedings of an international seminar. February 28-March 1, 2005. Westminister, London, UK. Crops Post harvest Programme (CPHP), Natural Resources International limited, Aylesford, UK.

Matches, R.J., 1982. Effects of temperature on the decomposition of pacific coast shrimp (*Pendalus jordani*). Journal of Food Science 47: 1065-1069.

Mendes, R., Huidobro, A. and Caballero, L.E., 2001. Indole levels in deepwater pink shrimp (*Parapenaeus longirostris*) from the Portuguese coast. Effects of temperature abuse. Europe Food Research and Technology 214: 125-130.

Moy, G., Kaferstein, F. and Motarjemi, Y., 1994. Application of HACCP to food manufacturing: some considerations on harmonization through training. Food Control 5: 131-139.

Nieto, B.G., Cinca, C.S. and Molinero, C.M., 2007. Microfinance institutions and efficiency. The International Journal of Management Science 35: 131-142.

Office Alimentaire et Veterinaire, 2002. Rapport concernant une mission en République du Bénin du 7 au 11 octobre 2002 concernant les conditions de production et d'exportation vers l'Union Européenne des produits de la pêche. Commission Européenne DG (SANCO)/8719/2002-MR Final. Available at: http://tinyurl.com/nogz8t3.

Office Alimentaire et Veterinaire, 2009. Rapport d'une mission effectuée au Bénin du 9 au 15 juin 2009 concernant les produits de la pêche. Commission Européenne DG (SANCO)/2009-8035-MR Final. Available at: http://tinyurl.com/pbj47j4.

Okoumassoun, L.E., Brochu, C., Deblois, C., Akponan, S., Marion, M., Averill-Bates, D. and Denizeau, F., 2002. Vitellogenin in tilapia male fishes exposed to organochlorine pesticides in Ouémé River in Republic of Benin. Science of the Total Environment 299: 163-172.

Pingali, P. and Khwaja, Y., 2005. Commercializing small farms: reducing transaction costs. Working papers 05-08, Agricultural and Development Economics Division of the Food and Agriculture Organization of the United Nations (FAO-ESA). Rome, Italy.

Pokharel, C., 2003. Agricultural diversification in Nepal. Paper presented at the international workshop on agricultural diversification and vertical integration in South Asia. November 5-7, 2003, New Delhi, India.

PSRSA, 2010. Plan Stratégique de Relance de du Secteur Agricole au Bénin: orientations stratégiques et plan d'action. MAEP, Benin.

Rapid alert system for food and feed (RASFF), 2011. Health and consumers. Annal report. Rapid alert system for food and feed. European Commission Publishers, Luxembourg, 52 pp. Available at: http://ec.europa.eu/food/food/rapidalert/index_en.htm.

République du Bénin, 2003a. Décret 2003/114 portant assurance qualité des produits de la pêche en république du Bénin. Official Journal 2003 No. 20.

République du Bénin, 2003b. Arreté 427/MAEP/D-CAB/SGM/DA/DP/CSRH/SA portant qualité des eaux utilisées dans les établissements à terre pour le traitement des produits halieutiques.

SCRP, 2007. Stratégie de croissance pour le réduction de la pauvreté. République du Bénin, Ministère de l'Agriculture, de l' Elevage et de la Pêche, Benin, 131 pp.

Soclo, H.H., Affokpon, A., Sagbo, A., Thomson, S., Budzinski, H., Garrigues, P., Matsuzawa, S. and Rababah, A., 2002. Urban run-off contribution to surface sediment accumulation of polycyclic aromatic hydrocarbons in the Cotonou lagoon, Benin. Polycyclic Aromatic Compounds 22: 111-128.

STDF, 2008. Overview of sanitary and phytosanitary needs and assistance in Benin. Standards and Trade Development Facility, 4 pp.

Van Greuning, H., Gallardo, J. and Randhawa, B., 1998. A framework for regulating microfinance institutions. financial sector development department. World Bank Economic Review 50. World Bank, Washington, DC, USA.

Weinberger, K. and Lumpkin, T.A., 2007. Diversification into horticulture and poverty reduction: a research agenda. World Development 35(8): 1464-1480.

World Bank, 2010. Benin: country brief. Available at: http://go.worldbank.org/RII10MVQJ0.

11. Quality management in supply chains of non-timber forest products: the case of gum arabic in Senegal[1]

G. Mujawamariya[1*], K. Burger[2] and M. D'Haese[3]

[1]Africa Rice Center, Dar es Salaam, P.O Box 33581, Dar es Salaam, Tanzania; [2]Wageningen University, Development Economics Group, Hollandseweg 1, 6706 KN Wageningen, the Netherlands; [3]Ghent University, Department of Agricultural Economics, Belgium, Coupure links 653, 9000 Ghent, Belgium; rgaudiose@gmail.com

Abstract

Low prices associated with variable quality of non-timber forest products in terms of users and consumers' needs and requirements are one of the factors limiting access and participation in markets. Quality can be determined on field or by the user. The current study explores the possibility to understand the current practices of producers in terms of quality supply and to link at least some of the users' quality criteria to production and marketing practices of producers. The study finds that good quality as defined on field is not always good when measured in laboratory; yet improving quality on field increases the likelihood of obtaining chemically good gum. Furthermore, determinants of supply by collectors and traders are investigated for two quality attributes namely size and cleanliness of gum nodules. Quality maintenance and improvement is influenced by harvest and post-harvest practices, behaviour and experience of traders, and price expectations. Research implications include the necessity for scholars interested in product quality to bring together the production and consumption sides because their perceptions and requirements may not always converge; regular trainings for collectors of non-timber forest products focusing on quality aspects; jointly establish clear rules of forest and market management in order to counteract the influence of market forces (price) on forests exploitation and enabling traders to have a definition of quality that is coherent and responsive to the actions and needs of collectors and users respectively.

Keywords: ordered logit, field assessment, laboratory assessment, marketing.

---

[1] Detailed description and analysis of results are found in Mujawamariya *et al.* (2012). Quality of gum arabic in Senegal: linking the laboratory research to the field assessment. Quarterly Journal of International Agriculture 51: 357-383. Authorisation to reproduce is granted by the journal publisher.

J. Bijman and V. Bitzer (eds.) **Quality and innovation in food chains**
DOI 10.3920/978-90-8686-825-4_11, © Wageningen Academic Publishers 2016

11.1 Introduction

Non-timber forest products (NTFPs) are recognised as important additional income sources in developing countries (Shackleton and Shackleton, 2004). The forest product activities including the harvest of non-timber forest products are undertaken to supplement the household incomes (Odebode, 2005) and to increase resilience in case of harvest failure. These activities act as coping strategies and provide seasonal employment (Kusters and Belcher, 2004; Ndione *et al.*, 2001; Prasad *et al.*, 1999). Although most non-timber forest products in Africa are freely accessed and collected, the collection of NTFPs yields poor returns and hence tends to provide only a basic level of income rather than substantially contributing to the integration of households in cash economies or providing a way of socio-economic advancement (Belcher *et al.*, 2005; Neumann and Hirsch, 2000; Prasad *et al.*, 1999).

Among the constraints hindering the adoption and harvest of the NTFPs are the lack of knowledge and technical support in the process of collection/production, as well as governance issues such as distribution of property rights to resources and the ability of local people to claim and enforce such rights (Ruiz Pérez and Byron, 1999). Factors limiting access to and participation in markets are also important. These include low quantities, low prices, lack of access to market information, remoteness of production zones and associated lack of proper infrastructure (especially transport, in case the product is perishable) and variable quality of the products in terms of the users and consumers' needs and requirements.

For most of the NTPFs, production areas are naturally endowed such that quantity is most readily available. However, one of the current challenges in the sector is that a high demand can lead to producers overexploiting the resource base; the demand can originate from the market or direct consumption needs of people (Lyndon, 2005). This overexploitation, especially when there are no clear management rules, causes resources degradation and thereby reduces future quantities in addition to impairing the quality of the harvested produce (Carr and Hartl, 2008; Rai, 2006; Sène and Ndione, 2007). Furthermore, the low product prices constrain collection and marketing of non-timber forest products (Marshall *et al.*, 2003). Low prices result from the underdeveloped marketing system and the importance of interlocked exchanges (Mujawamariya and D'Haese, 2012), but also from the lack of knowledge of product quality by producers (Ahenkan and Boon, 2010). According to Prasad *et al.* (1999), collectors are sometimes cheated and offered low prices by traders on pretexts of impurities, moisture or inferior quality.

The lack of knowledge on product quality does not mean that collectors' traditional knowledge is redundant, rather it shows a certain disconnect between producers' practices and users' requirements particularly when the products are traded in international markets. Some studies recommend to trade NTFPs exclusively in local markets where familiarity guarantees acceptability whereas in export markets users' preferences are not known (De Beer and McDermott, 1996; Kusters and Belcher,

2004; Shackleton *et al.*, 2007). In exports markets, quality standards are high and lack of information by rural producers on how to contact (and make contracts) with users and on the required product quantity and quality consistency further limits market access (Marshall *et al.*, 2003, Shanley *et al.*, 2008). Consequently, success of NTFPs production and marketing depends on factors which are intertwined and all are to some extent related to quality. Despite its importance, quality aspects are not sufficiently dealt with in the NTFP literature. A few studies indicated that quality is determined by the genetic variability and environment (Schmidt, 2005), mode and time of harvest (Marshall *et al.*, 2003; Ahenkan and Boon, 2010), post-harvesting handling (Lamien *et al.*, 1996), and market demand (Mander, 1998). However, the above factors were not empirically studied.

Against this background, this study aims to contribute to the understanding of quality aspects of NTFPs through an empirical analysis which considers the quality criteria and determinants of quality supply. In the first stage, we explore criteria that determine the choice of the users (in our case industry) including the specific properties of the product and compare these criteria to the quality assessment of the product by producers with the aim of testing the familiarity of the producers with the consumer preferences. In the second stage, we analyse the influence of the determinants of quality on its supply in the market; these determinants include factors associated with harvest, post-harvest handling, environment and market.

We take gum arabic production in Senegal as our case study. Gum arabic is a non-timber forest produced by *Acacia senegal* species commonly found in the semi-arid lands of Africa. It is the oldest and the best known of all the natural gums; its uses are dated to about 4,000 years BC in inks, paintings, cosmetics, clothing, medicine and mummification process (Alland, 1944). It was also consumed as food. Currently, gum arabic is widely used in food and non-food industries where it functions for instance as an adhesive, emulsifier, stabiliser, thickener, flavouring or coating agent, etc. (Glicksman, 1969; Wickens *et al.*, 1995, Williams and Phillips, 2009). It is also used in non-food industries for instance in modern pharmacy (Khan and Abourashed, 2010). Ali *et al.* (2009) suggest a possible use of gum arabic in dentistry because it enhances dental remineralisation and has some antimicrobial activity. Other commercial uses of gum arabic are found in ink production, pottery pigments and glazing for colour thickening in water-colours and paints, wax polish or for giving lustre to silk and crepe in textiles and lithography (Wickens *et al.*, 1995). Depending on the uses of gum arabic, which are very diverse, different quality grades of gum are needed. Typically the food or pharmaceutical industries would require the finest quality whereas the painting or textile industries would not necessarily request the best quality (Agrigum, 2011; Somo, 2009).

Quality of gum arabic is usually assessed by the user based on some intrinsic quality attributes, including physicochemical properties, such as protein content, ash content, specific rotation and average molecular mass (Anderson *et al.*, 1966; Williams and Phillips, 2009). Producers on the other hand, with no knowledge of these sophisticated measurements, assess gum quality based on a visual inspection of the exterior attributes of gum including colour, cleanliness or size of the nodules. At times, however, there is discontent on the part of users who reject gum or offer lower prices on the basis of low quality supply. This study is therefore pertinent as it explores the possibility to understand the current practices of producers in terms of quality supply and to link at least some of the users' quality criteria to production and marketing practices of producers so that such rejections could be minimised in the future.

The following section provides an overview of the literature on quality aspects specially related to the food sector in general, and to gum arabic in particular. Section 11.3 describes the methodology while in Section 11.4 we present results of the empirical study. Section 11.5 includes the discussion, results and conclusions, and suggestions for some practical implications and opportunities for future research.

11.2 Quality aspects and gum arabic quality

Quality is generally defined as a measure of excellence. It is a widely used concept but it often remains abstract and complex. In production, quality is a state of being free from defects and significant variations, brought about by a strict and consistent adherence to measurable and verifiable standards with the purpose of achieving uniformity of output that satisfies specific user requirements (Dale *et al.*, 1997). In business, quality of goods and services refers to the creation of customer satisfaction and is one of the elements that contribute to profitability (Evans and Lindsay, 2005).

Quality has several attributes including intrinsic and extrinsic quality attributes. According to Luning and Marcelis (2009), intrinsic quality attributes are objectively and physically linked to the product while extrinsic quality attributes do not necessarily have a direct relationship with the product properties. These attributes can however affect the users' quality perception or the product's acceptance. Examples of intrinsic attributes include texture, taste, protein content or microbial condition. Examples of extrinsic attributes include religious rules, organic production, or brand name which can subjectively influence the consumers' quality perception or product acceptance. According to Swinnen and Maertens (2007), these aspects are most pronounced in western markets as well as in middle-class urban markets in low-income countries, often under high quality and safety demands.

Quality aspects are often associated with a cost and a price. The cost of quality means the cost undertaken in the process of improvement including efforts to prevent quality deterioration during handling, transport or storage (Campanella, 1999). The price for quality is a premium such that differences in prices of the same product indicate quality differences (Tollens and Gilbert, 2003). According to Fafchamps *et al.* (2008), these differences can even be translated into well-defined grades. Through grading, producers can target high-value markets with their highest quality produce, and generate higher returns on investments (Evans and Lindsay, 2005).

Similar to other products, gum quality possesses both intrinsic and extrinsic attributes. Intrinsic attributes of gum arabic are visible or invisible. They are associated with active structural and physical properties. Visible attributes are size and fullness or hardness of its nodules, colour and cleanliness. They can be changed intentionally (e.g. through post-harvest cleaning) or unintentionally through the product's interaction with the environment (e.g. as the gum matures or dries, it can change colour). Invisible attributes include gum's chemical composition, which is permanent (Li *et al.*, 2012). Extrinsic attributes of gum arabic include for instance being organically or sustainably collected (this is not labelled in the case study markets).

Gum arabic is not directly consumed; it is an ingredient in food and non-food processing. Hence in its industrial application, the extrinsic attributes are less important than the intrinsic quality aspects. The visible quality attributes are not only simple indicators of quality applicable under field conditions, they are also important for grading gum by collectors and traders. Grading is important because it helps targeting markets which are undersupplied (Marshall *et al.*, 2003). Senegal and other small gum producing countries do not have a known grading system through which quality can be established. Sudan, Nigeria and Chad have registered their gum grades on the world markets such that these grades have become internationally known. Table 11.1 presents the grading system of Sudan.

As these grades command different prices, the absence of a grading system for a producing country is a serious problem. It implies the loss of not only the opportunity to transact in specific markets but also of price premiums that are associated with superior grades.

Table 11.1. Sudan classification of gum arabic (Macrae and Merlin, 2002; Williams and Phillips, 2009).

Grade	Description[1]	% at sorting
Handpicked selected	Cleanest, lightest colour, and whole nodule, Ø >30 mm; most expensive grade	0 to 5
Cleaned amber and sifted	Clean and siftings are removed, pale to dark amber colour, whole or broken nodule, Ø >20 mm	5-10
Cleaned	Standard grade, contains siftings but dust is removed, whole nodule plus fragments, 10< Ø <20 mm	70
Siftings	Fine particles left after sorting, contains sand, bark and dirt, 2.5< Ø <10 mm	5
Dust	Very fine particles collected after the cleaning process, Ø <2.5 mm	5
Red gum	Dark and red particles, only for local use	

[1] Ø = diameter.

Within *Acacia senegal* species, the main factors affecting grades of gum arabic quality are different botanical sources (varieties), tapping methods and harvesting period and environmental factors (Chikamai and Odera, 2002). First, the known varieties of *Acacia senegal* include var. *senegal,* var. *kerensis*, var. *rostrata* and var. *leiorhachis*. In the zones of study in Senegal and also in Sudan, only the var. *senegal* grows (Cossalter, 1991). Second, tapping is commonly used in Senegal and Sudan, it is carried out during the early part of the dry season during October-November (Phillips, 2012). The modern tapping tool used in Sudan is a sharp spear (sonki). In Senegal the axe is also commonly used besides the sonki (Idris and Haddad, 2012; Okatahi and Onyibe, 1999.). It is suggested that gum should be left to mature and form nodules for at least 14 days after which it can be harvested off the tree (Dione and Sall, 1988). Depending on the area, this period can even be extended to 4 or 6 weeks (Idris and Haddad, 2012). The long maturity period poses the risk of appropriation of gum by competing collectors particularly in communal forests where access to the common plots is unrestricted and the rules of management are unclear (Sène and Ndione, 2007). In these common plots, increased competition leads to increased levels of extraction of the resource before maturity, thereby leading to low quality. The nature of resource ownership and use is indicated as the main raison behind competition: in private plots and community-managed forests where competition is regulated, quality is better and in open-access forests where the 'first come, first serve' rule operates, insecure tenure over collection areas leads to risk of over-exploitation and inability to manage quality (Belcher and Schreckenberg, 2007).

Third, to maintain gum quality during gum collection, collectors need to apply appropriate tapping techniques, tap at the right time (following climatic and ecological indications), and respect the waiting period after tapping. They should harvest gum off the tree carefully without taking it with the tree bark, and prevent it from falling on the soil. Post-harvesting handling is also important: cleaning, proper drying and storage improve the quality of gum arabic (Okatahi and Onyibe, 1999; Phillips, 2012). Ramly (2012) emphasised the benefits of improving the technical know-how in the treatments and activities associated with gum production even if collectors generally learn the above-mentioned good practices through contact with other collectors. Finally, environmental factors are important in the production of gum arabic: Wekesa *et al.* (2010) and Ballal *et al.* (2005) found that rocky soils, high soil temperatures and high rainfall lead to better yields.

Problems with quality are also related to improper handling of gum arabic, such as mixing it with other types of gums to increase weight, harvesting immaturely to get it off the tree before other collectors, not cleaning of its impurities, drying or storing improperly and not grading (DEFCCS, 2005; Seif el Din and Zarroug, 1996). Such problems make that quality of gum supplied is uncertain unless the collector or any subsequent buyer takes a deliberate effort to clean and grade it. Variations in quality have also led to design of artificial gum substitutes that have enhanced and reliable properties (Aoki, *et al.*, 2007). Therefore, gum producing countries, including Senegal, need to be aware of maintaining and improving the quality of the produced gum in order to retain and expand their market share.

11.3 Methodology

Two methodological approaches are followed. First, a comparison was conducted between the assessment of quality and chemical analyses of gum arabic, which were done in the laboratory. Second, determinants of quality supply were examined in the context of current production and marketing practices.

11.3.1 Assessment of quality

On the one hand, a field assessment was done during the visit of 11 villages from the Sylvopastoral zone and Eastern region of Senegal between March and April 2010. In total, 27 samples of gum were obtained; these samples were randomly picked from sacks of gum arabic intended for sale. They were assessed and graded by collectors and traders in the village. Samples were classified as of best, first, second or third quality on the basis of intrinsic visible quality attributes including the size of the gum nodule and cleanliness.

Secondly, an analysis of chemical components which determine gum properties and thereby its usage in different industries was done in the laboratory. Chemical analyses were done to determine intrinsic invisible quality attributes of all the 27 gum samples which were previously assessed on the field. The laboratory analyses include measurements of mineral matter or ashes (determined on gum powder by combustion at 55 °C during 4 h), and the specific rotatory power (determined by direct reading of rotator power on a Bellingham + Stanley's ADP220 polarimeter).

Results from chemical analyses were compared with the field assessments to determine the extent to which the samples fulfilled the users' requirements. In order to have the margins of acceptability of gum arabic based on chemical measurements, two representative specifications of users provided the basis for comparison, namely the JEFCA for food products in general at the international level and Valdafrique[2] for pharmaceutical products at the national level.

11.3.2 Analysis of quality supply by gum collectors

In the second step of the analysis, the determinants of supply of quality are analysed for two visible attributes namely the size and cleanliness of the nodules: the bigger and/or the cleaner the gum nodule, the higher is its quality. The variable indicating

---

[2] Valdafrique operates in Senegal since 1943. It specialises in the processing of gum into final products, such as medical tablets. It also transforms gum into a spry-dried gum powder which is used as a semi-raw product by the gum-using industries.

the size of gum nodules is defined by the proportion in the quantity brought by the collector to the market: a large proportion of nodules that are smaller than 2 cm (coded as 0), a small proportion of nodules that are smaller than 2 cm (coded as 1), a small proportion of nodules that are larger than 2 cm (coded as 2), and a large proportion of nodules that are larger than 2 cm (coded as 3). The variable indicating the cleanliness of nodules is defined by the proportion in the quantity brought by the collector to the market: a small proportion of nodules that are clean (coded as 0), about half of the nodules that are clean (coded as 1), and a large proportion of nodules that are clean (coded as 2). These values indicate attributes of the ordered observed outcome in the supply of quality.

Following Dione and Sall (1988), Chikamai and Odera (2002), and Idris and Haddad (2012), determinants of quality include tapping aspects, post-harvest handling and environmental factors. These determinants are operationalised by the duration between tapping the *Acacia senegal* tree and collecting the gum off the tree, type of forest management, participation in training on quality and time spent in post-harvest activities. We add control variables for environmental factors proxied by seasonal differences (which for instance imply differences in humidity, temperatures) and regional differences (which for instance express differences in soil characteristics and rainfall). Furthermore, we also add market factors because they also influence the supply of gum quality. Here we include the place where gum is sold, trader experience, and price in previous season.

Data was obtained through a system of market monitoring done between January and May 2010 with the purpose of recording the quality brought to the markets together with the specific details that explain the supply of quality. This monitoring was conducted in 16 markets where farmers sell their produce. A sample of collectors who were interviewed during the monitoring process was randomly constituted in a quota of 10 collectors per market visit. The markets were visited three times in the season (beginning, mid and end of the collection season) resulting in 219 formal records (i.e. a 60% response rate). Table 11.2 provides an overview of the descriptive statistics of the variables used as determinants of quality supply in terms of size of gum nodules and cleanliness.

Following Greene (2008), the analysis of determinants of quality aspects is done through an ordered logit model (ologit). The use of the ologit model is justified by nature of the dependent variables. Indeed, the observed attributes of the size or cleanliness of gum nodules which are the dependent variables indicate ordered categories of quality from low to high quality.

Table 11.2. Descriptive statistics of determinants of quality supply.

Variable	Size of nodules					Cleanliness		
	All (219)	Many <2 cm (89)	Few <2 cm (74)	Few >2 cm (46)	Many >2 cm (10)	Few clean (19)	Half clean (136)	Many clean (64)
Distribution of quality attributes		0.41 (0.03)	0.34 (0.03)	0.21 (0.03)	0.04 (0.01)	0.09 (0.02)	0.62 (0.03)	0.29 (0.03)
Duration between tap and collection (1: ≥14 days)	0.33 (0.47)	0.32 (0.47)	0.30 (0.46)	0.39 (0.49)	0.22 (0.44)	0.23 (0.43)	0.33 (0.49)	0.37 (0.48)
Forest management (1: commune)	0.52 (0.50)	0.44 (0.53)	0.37 (0.49)	0.67 (0.47)	0.48 (0.50)	0.61 (0.50)	0.41 (0.49)	0.73 (0.44)
Post-harvest time (minute/quantity)	5.83 (6.09)	4.65 (4.64)	7.14 (7.77)	5.76 (5.61)	6.82 (3.43)	5.82 (4.72)	4.68 (4.19)	8.27 (8.69)
Training (1: participated)	0.72 (0.45)	0.44 (0.53)	0.75 (0.43)	0.63 (0.49)	0.76 (0.43)	0.61 (0.50)	0.71 (0.45)	0.75 (0.17)
Sale place choice (1: village boutique)	0.56 (0.50)	0.45 (0.50)	0.58 (0.50)	0.76 (0.43)	0.44 (0.53)	1	0.54 (0.50)	0.48 (0.50)
Trader experience (year)	15.6 (9.39)	17.7 (9.10)	12.3 (10.30)	17.2 (7.76)	13.2 (1.72)	17.2 (6.91)	18.1 (8.73)	9.9 (8.98)
Price in previous season (CFA/kg)	616.0 (158.82)	584.8 (133.87)	711.5 (171.33)	543.6 (116.38)	516.7 (86.60)	560.5 (149.61)	564.3 (124.56)	742.2 (157.41)
Begin-season (1: if collection at the beginning)	0.30 (0.46)	0.61 (0.49)	0.15 (0.36)	0	0	0.06 (0.24)	0.37 (0.48)	0.22 (0.42)
Mid-season (1: if collection in mid-season)	0.32 (0.47)	0.19 (0.39)	0.38 (0.49)	0.43 (0.50)	0.67 (0.50)	0.5 (0.51)	0.3 (0.48)	0.22 (0.42)
End of season (1: if collection at end)	0.38 (0.48)	0.20 (0.40)	0.47 (0.50)	0.56 (0.50)	0.33 (0.50)	0.44 (0.51)	0.28 (0.45)	0.56 (0.50)
Region (1: Eastern Senegal)		0.13 (0.04)	0.62 (0.06)	0.11 (0.05)	0	0.17 (0.09)	0.10 (0.02)	0.73 (0.05)

11.4 Results

11.4.1 Assessment of quality

The classification of samples according to the field criteria and results from chemical analyses are presented in Table 11.3.

Collectors classified the 27 samples in three grades: 16 samples as best, first, or standard quality grade, 8 samples as second quality grade and 3 samples as third quality grade. Criteria of good quality in the field are that gum nodules are large (diameter >2 cm) and clean (Macrae and Merlin, 2002). According to these criteria, more than 50% of the samples were of good quality. For food and chemical industries, good quality means that the ash content should not exceed 4% of dry matter and optical rotation should be between -26° and -34° (JEFCA, 2006; Valdafrique, 2011)[3].

---

[3] Details of assessment are in Appendix 11.1.

Table 11.3. Quality assessment of 27 gum arabic samples.

Quality criteria		Number of samples			Total as percentage
		1st grade (16)	2nd grade (8)	3rd grade (3)	
Size	small pieces	7	4	2	48
	large pieces	9	4	1	52
Cleanliness	dirty	3	6	2	41
	clean	13	2	1	59
Ash content	less than 4%	10	2	1	48
	more than 4%	6	6	2	52
Optical rotation	<-34°	2	0	1	11
	-34° to -26°	12	7	2	78
	>-26°	2	1	0	11

According to the industrial requirements, less than 50% of the samples meet the ash content criterion and 78% of the samples have the required optical rotation. In comparing the laboratory assessment to the field assessment, only 8 samples among the 16 first grade samples were correctly found to fulfil the optical rotation and ash content criteria jointly. Among the samples of low quality as assessed on the field, 2 samples from the 2nd grade and 3rd grade each were incorrectly classified by collectors as they were found to be of good quality in accordance with the laboratory analyses.

11.4.2 Determinants of quality supply by gum collectors

The analysis of determinants of quality supply by gum collectors is done through an ordered logit. Table 11.4 shows the results for the two quality attributes namely (1) the nodule size (in four categories: a large proportion of nodules that are smaller than 2 cm, a small proportion of nodules that are smaller than 2 cm, a small proportion of nodules that are larger than 2 cm, and a large proportion of nodules that are larger than 2 cm) and (2) cleanliness of gum nodules (in three categories: a small proportion of nodules that are clean, about half of the nodules that are clean, and a large proportion of nodules that are clean.

The results of the models of quality supply with respect to the size of gum nodules and cleanliness of gum nodules show that the large gum nodules are supplied by collectors who obey enough time for the gum to mature between tree tapping and gum harvesting (at least 14 days), spend long time on post-harvest activities, participate in training, and choose to sell to the village boutique. In comparison with the beginning of the season, larger gum nodules are obtained during the mid-season or towards the end of season of collecting and selling the gum. Collection in communal forests in comparison with privately owned plots has a negative effect on the size of nodule; such competition increases due to a high price in previous season, hence the latter also has

Table 11.4. Ordered logistic results (nodule size: 4 categories; nodule cleanliness 3 categories).

Variables	Size of nodules[1,2]	Cleanliness of nodules[1,2]
Duration between tap and harvest (1: ≥14 days)	Positive[+]	Positive
Forest management (1: commune)	Negative[+]	Negative
Post-harvest time (In minute/quantity)	Positive*	Positive*
Training (1: participated)	Positive**	Positive
Sale place choice (1: village boutique)	Positive***	Negative[+]
Trader experience (year)	Negative *	Positive [+]
Price in previous season (CFA/kg)	Negative *	Negative *
Mid-season (1: if collection in mid-season)	Positive ***	Negative [+]
End of season (1: if collection at end)	Positive***	Positive
Region (1: Eastern Senegal)	Negative	Positive***
Cut-off1	0.736 (1.192)	-0.976 (1.317)
Cut-off2	2.819 (1.204)	3.141 (1.343)
Cut-off3	5.122 (1.228)	
Log-likelihood	-208.136	-148.707
LR-chi square	106.70***	82.53***
Pseudo R-square	0.204	0.217
% good predictions	53.88	78.08

[1] Standard errors between parentheses.

[2] *** significant at 1% level, ** significant at 5% level; * significant at 10% level, + significant at 15% level.

a negative effect on quality. Moreover, traders who are new in gum business prefer large nodules. On the other hand, the clean gum nodules are supplied by collectors who wait long enough for gum maturation between tapping and collecting of gum, and spend long time on post-harvest activities. Experienced traders prefer cleaner nodules whereas selling to the village boutique and a high price in the previous season have a negative effect on cleanliness. Furthermore, in comparison with the beginning of the season, the gum collected during the mid-season is less clean. Eastern Senegal is the region where cleaner gum is mostly produced.

11.5 Discussion and conclusions

By comparing the field and laboratory assessment of gum quality to the users requirements and specifications in the international food and local pharmaceutical industries, we found that low quality in the field is in most cases confirmed to be bad by any technical assessment and high quality from the perspective of collectors and traders is not always good in terms of laboratory assessment. More samples were considered to be of low quality by the laboratory tests compared to visual inspection. This confirms the views of Schmidt (2005) who emphasises that quality is not easy to 'see' and strongly supports the use of physicochemical analyses to ensure quality especially when products are valuable ingredients. Moreover, this divergence in quality measurements influences rejection rates because the evaluation of these physicochemical characteristics is commonly made during trade at the moment of export (Okullo, 2010). The results also leads to an interesting observation that if supply of high quality in the field could be increased, it would be more likely to fulfil quality requirements in terms of laboratory assessment. At this point it is important to note that even though many of the samples are labelled as 'bad', this does not mean that they are useless. Rather, they may be used in other industries, for instance in painting or textile, where quality requirements are less stringent (Derrick *et al.*, 1999). Therefore, it is important to recognise different grades, and to target industrial niches for each grade.

Harvest and post-harvest handling had a positive influence on supply of quality: the long duration between tapping and harvesting gum and the more time spent on post-harvest practices (including cleaning and sorting) lead to better gum. According to Prasad *et al.* (1999), incomes could be significantly raised by such simple value addition practices which would increase the remuneration to the collectors. This implies that there is need for more attention for the basics of the techniques of gum harvesting and practices to maintain and improve its quality. Collectors can acquire the related knowledge and skills through experience, but trainings are needed specially to link the collector's knowledge to the quality aspects that are important to the users; the training conducted at the beginning of the season proved to be important in that respect.

The behaviour of a village boutique owner is not consistent with respect to quality attributes. The village boutique owner (commonly called as *boutiquier*) are established in gum producing villages and buy gum throughout the harvesting season. They are distinguished from mobile traders known as '*banabana*'. These mobile traders buy gum in producing villages, directly from collectors or from village boutique owners. They also participate in rural weekly markets or towns neighbouring the production zones. In terms of quality, the village boutique owner is interested in the big nodules of which inspection is quick at low cost and associated with quantity, however he ends up with the less clean gum. For the trader, an emphasis on cleanliness not only reduces the quantity to trade but also leads to a problem of enforcing the quality requirements as it is difficult and costly to reject gum: repeated rejections of the supplied gum results in a negative social outcome and collectors will not go to him in the future which means that he can lose a market altogether. On the other hand, the less experienced trader who would like to establish himself in business and needs to build a large supply base does not strongly enforce quality requirements. These results are reversed with regard to the trader's experience in the gum business, allowing us to safely assume that village boutique owners are younger traders who are interested in larger nodules compared to the older traders. Proper assessment of cleanliness is achieved by experience; hence the experienced traders are indeed more interested in clean gum. The consequence of erratic behaviour on the side of traders is that the supplied quality remains low. This is because instead of rewarding high quality, traders do not penalise poor quality but accept it implying that the market mechanism to reward high quality production does not work. Therefore, gum arabic production remains in a low equilibrium trap whereby the quality traded remains of low quality because it is not penalised by a lower price and the high quality is not rewarded by a higher price.

The effect of price is discussed in the forest management context. It is generally expected that higher prices generate incentives to upgrade quality; however, this is only possible if competition in common forests is controlled. In case of unregulated competition, high prices in previous seasons lead collectors to expect high prices on the market and increase the quantity of supply as collectors want to increase their revenues. This focus on quantity becomes detrimental to quality: gum is not left to mature enough but is picked when the nodule is still small and has not sufficiently dried. Such gum is often found with impurities (DEFCCS, 2005; Seif el Din and Zarroug, 1996). Cases of declining quality due to rising prices have been observed in other non-timber products especially in open-access forests (Belcher and Schreckenberg, 2007; Lamien *et al.*, 1996; Neumann and Hirsch, 2000).

With regard to environmental factors, there are indeed significant differences within the season and across regions. These differences are associated with humidity, wind, heat, rainfall or soil which affect gum quality as also specified by Chikamai and Odera (2002). Taking the beginning of season as the base, we found that the size of nodules supplied increases throughout the season and cleanliness declines in mid-season to increase again at the end of season.

11.6 Recommendations for practice and further research

While the above results cannot be generalised to all the non-timber forest products, the attempt to translate the users' quality criteria into the production and marketing practices of producers by linking the laboratory measurements to the field assessment and then investigating determinants of quality has led to several interesting findings and practical implications. First, it has become evident that it is indeed useful for scholars interested in product quality to bring together the production and consumption sides because their perceptions and requirements may not always converge: the objectively measureable quality is not always directly linked to visible quality attributes in the field. Knowledge of quality supplied on the confirmed criterion such as cleanliness helps in the process of targeting market niches or reducing risk of rejection by the buyer.

Secondly, supply of good quality in the field increases the likelihood of obtaining good quality produce on the basis of invisible attributes. Hence, collectors should be sensitised to put efforts in respecting good harvest and post-harvest practices. Training of collectors of non-timber forest products focusing on quality aspects should be regularly conducted as knowledge and awareness creation are probably the main building blocks for quality-oriented production. The above trainings should build on the collectors' traditional knowledge and extend to the quality-improvement aspects that they can easily relate with and assimilate.

Thirdly, the current study has highlighted the need to better understand the role of forest management on quality. Clear rules of forest and market management are jointly needed to regulate the influence of market forces (price) on forests exploitation.

Fourthly, including specific market factors in the study of quality revealed that traders' behaviour may rather be uncertain towards quality. As intermediaries between the collectors and users, traders should have a definition of quality that is coherent and responsive to the actions and needs of collectors and users respectively. Hence it is important that traders are also targeted in training. A common understanding will reduce the quality uncertainties which also negatively affect the price that traders pay. Traders could play an important role in the search and transmission of information on quality.

There are several aspects which deserve further research including an investigation of the influence of specific environmental factors on quality. Proxies for such environmental factors were included in the study but it is necessary to include real measures so that collectors can distinguish clear influences and anticipate quality changes between and during seasons. Such studies would be beneficial in terms of determining the consistency of quality. Furthermore, there is need to understand the determinants of demand and supply of differentiated quality. As the next step of linking quality practices to users' specifications, it is important to investigate the requirements for implementing certification for gum arabic in particular and other non-timber forest products in general.

Acknowledgements

We are grateful to the European Union for funding the INCO-DEV ACACIAGUM Project through which this research was conducted. We are also thanking Dr Julia Wilson of CEH, UK for sending the gum samples to the CEH and CIRAD laboratories where chemical analyses were conducted; we extend our thanks to these laboratories.

References

Agrigum International, 2011. Gum acacia natural products worldwide. Available at: http://www.agrigum.com/docs/Corporate_Brochure.pdf.

Ahenkan, A. and Boon, E., 2010. Commercialization of non-timber forest products in Ghana: processing, packaging and marketing. Journal of Food, Agriculture and Environment 8: 962-969.

Ali, B.H., Ziada, A. and Blunden, G., 2009. Biological effects of gum Arabic: a review of some recent research. Food and Chemical Toxicology 47: 1-8.

Alland, R., 1944. Les gommes industrielles. Cours conférences 1310.

Anderson, D.M.W., Sir Edmond Hirst and Stoddart, J.F., 1966. Some structural features of Acacia Senegal gum (gum Arabic). Journal of the Chemical Society C: 1959-1966.

Aoki, H., Al-Assaf, S., Katayama, T. and Phillips, G.O., 2007. Characterization and properties of *Acacia senegal* (L.) Willd. var. senegal with enhanced properties (*Acacia* (sen) Super Gum™). Mechanism of the maturation process. Food Hydrocolloids 21: 329-337.

Ballal, M.E., El Siddig, E.A., Elfadl, M.A. and Luukkanen, O., 2005. Relationship between environmental factors, tapping dates, tapping intensity and gum Arabic yield of an Acacia Senegal plantation in western Sudan. Journal of Arid Environments 63: 379-389.

Belcher, B., Ruiz-Pérez, M. and Achdiawan, R., 2005. Global patterns and trends in the use and management of commercial NTFPs: implications for livelihoods and conservation. World Development 33: 1435-1452.

Belcher, B. and Schreckenberg, K., 2007. Commercialization of non-timber forest products: a reality check. Development Policy Review 25: 355-377.

Campanella, J., 1999. Principles of quality costs: principles, implementation, and use. ASQ Quality Press, Milwaukee, WI, USA.

Carr, M. and Hartl, M., 2008. Gender and non-timber forest products – Promoting food security and economic empowerment. International Fund for Agricultural Development, Rome, Italy.

Chikamai, B.N. and Odera, J.A., 2002. Commercial plant gums and gum resins in Kenya: sources of alternative livelihood and economic development in the drylands. Executive Printers, Nairobi, Kenya.

Cossalter, C., 1991. *Acacia senegal*, gum tree with promise for agroforestry. Nitrogen Fixing Tree Association, Waimanalo, HI, USA.

Dale, B.G., Williams, A.R.T., Barber, K.D. and Van Der Wiele, A., 1997. Managing quality in manufacturing versus services: a comparative analysis. Management Service Quality 7: 242-247.

De Beer, J.H. and Mc Dermott, M.J., 1996. The economic value of non-timber forest products in Southeast Asia. IUCN, Amsterdam, the Netherlands.

Derrick, M.R., Stulik, D. and Landry, J.M., 1999. Infrared spectroscopy in conservation science – Scientific tools for conservation. The Getty Conservation Institute, Los Angeles, CA, USA.

Direction des Eaux, Forêts, Chasses et de la Conservation des Sols (DEFCCS), 2005. Etude diagnostique de la filière des gommes et résines et plan décennal de développement du projet. Projet Opération Acacia. Dakar, Sénégal.

Dione, M. and Sall, P.N., 1988. Le gommier et la gomme arabique au Sénégal. Bilan des actions de recherche et de développement. Perspectives d'avenir. Troisième symposium sur le gommier et la gomme arabique. DRPF, Saint Louis, MO, USA.

Evans, J.R. and Lindsay, W.M., 2005. The management and control of quality. Thomson South-Western, Mason, IL, USA.

Fafchamps, M., Hill, R.V. and Minten, B., 2008. Quality control in non-staple food markets: evidence from India. Agricultural Economics 38: 251-266.

Glicksman, M., 1969. Gum technology in the food industry. Food Science and Technology 8. Academic Press, New York, NY, USA.

Greene, W.H., 2008. Econometric analysis. Pearson-Prentice Hall, Upper Saddle River, NJ, USA.

Idris, O.H.M. and Haddad, G.M., 2012. Gum Arabic's (gum acacia's) journey from tree to end user. In: Kennedy, J.F., Phillips, G.O. and Williams, P.A. (eds.) Gum Arabic. Royal Society of Chemistry, Cambridge, UK, pp. 3-18.

Joint FAO/WHO Expert Committee on Food Additives (JECFA), 2006. Combined compendium of food additive specifications, analytical methods, test procedures and laboratory solutions used by and referenced in the food additive specifications. Monographs 1(4). FAO, Rome, Italy.

Khan, I.E. and Abourashed, E.A., 2010. Leung's encyclopaedia of common natural ingredients used in food, drugs and cosmetics. John Wiley and Sons, New York, NY, USA.

Kusters, K. and Belcher, B., 2004. Forest products, livelihoods and conservation. Case studies of non-timber forest products. CIFOR, Bogor, Indonesia.

Lamien, N., Sidibe, A. and Bayala, J., 1996. Use and commercialization of non-timber forest products in western Burkina Faso. In: Leakey, R.R.B., Temu, A.B., Melnyk, M. and Vantomme, P. (eds.) Domestication and commercialization of non-timber forest products in agroforestry systems. Non-wood forest products 9. FAO, Rome, Italy, pp. 51-64.

Li, X., Zhang, H., Fang, Y., Al-Assaf, S., Phillips, G.O. and Nishinari, K., 2012. Rheological properties of gum Arabic solution: the effect of Arabinogalactan Protein Complex (AGP). In: Kennedy, J.F., Phillips, G.O. and Williams, P.A. (eds.) Gum Arabic. Royal Society of Chemistry, Cambridge, UK, pp. 229-238.

Luning, P.A. and Marcelis, W.J., 2009. Food quality management: techno-managerial principles and practices. Wageningen Academic Publishers, Wageningen, the Netherlands.

Lyndon, J., 2005. The potential of non-timber forest products to contribute to rural livelihoods in the windward islands of the Caribbean. CANARI Technical Report No. 334. Caribbean Natural Resources Institute, Laventille, Trinidad.

Macrae, J. and Merlin, G., 2002. The prospects and constraints of development of gum Arabic in sub-Saharan Africa. A document based on the available literature and field trips to Chad, Mali and Niger. Africa Tree Crops Initiative. World Bank, Washington, DC, USA.

Mander, M., 1998. Marketing of indigenous medicinal plants in South Africa. FAO, Rome, Italy.

Marshall, E., Newton, A.C. and Schreckenberg, K., 2003. Commercialization of non-timber forest products: first steps in analysing factors influencing success. International Forestry Review 5: 128-137.

Mujawamariya, G. and D'Haese, M., 2012. In search for incentives to gum Arabic collection and marketing in Senegal: interlocking gum trade with pre-finances from traders. Forest Policy and Economics 25: 72-82.

Ndione, C.M, Sène, A., Dieng, A.B., Diop, O., 2001. Etude de l'organisation et des performances de filières forestières. BAME, Dakar, Sénégal.

Neumann, R.P. and Hirsch, E., 2000. Commercialization of non-timber forest products: review and analysis of research, CIFOR, Bogor, Indonesia.

Odebode, S.O., 2005. Contributions of selected non-timber forest products to household food security in Nigeria. Journal of Food, Agriculture and Environment 3: 138-141.

Okatahi, S.S. and Onyibe, J.E., 1999. Production of gum Arabic. National Agricultural Extension and Research Liaison Services. Extension Bulletin 78. ABU, Zaria, Nigeria.

Okullo, J.B., 2010. Physico-chemical characteristics of shea butter (Vitellariaparadoxa C.F. Gaertn.) oil from the shea districts of Uganda. African Journal of Food, Agriculture, Nutrition and Development 10: 2070-2084.

Phillips, G.O., 2012. Preface. In: Kennedy J.F., Phillips, G.O. and Williams, P.A. (eds.) Gum Arabic. Royal Society of Chemistry, Cambridge, UK.

Prasad, R., Das, S. and Sinha, S., 1999. Value addition options for non-timber forest products at primary collector's level. International Forestry Review 1: 17-21.

Rai, N.D., 2006. The socio-economic and ecological impact of Garcinia gummi-gutta fruit in the Western Ghats, India. In: Kusters, K. and Belcher, B. (eds.) Forest products, livelihoods and conservation. Case studies of non-timber forest products. CIFOR, Bogor, Indonesia.

Ramly, F.M.A., 2012. Participatory forest management: the role of farmers associations in the rehabilitation of the gum Arabic belt. In: Kennedy J.F., Phillips, G.O. and Williams, P.A. (eds.) Gum Arabic. Royal Society of Chemistry, Cambridge, UK, pp. 81-86.

Ruiz-Pérez, M. and Byron, N., 1999. A methodology to analyse divergent case studies of non-timber forest products and their development potential. Forest Science 45: 1-14.

Schmidt, K., 2005. NTFPs and poverty alleviation in Kyrgyzstan: potential and critical issues. In: Pfund, J.-L. and Robinson, P. (eds.) Non-timber forest products between poverty alleviation and market forces. Inter cooperation, Bern, Switzerland, pp. 28-29.

Seif El Din, A.G. and Zarroug, M., 1996. Production and commercialization of gum Arabic in Sudan. In: Leakey, R.R.B., Temu, A.B., Melnyk, M. and Vantomme, P. (eds.) Domestication and commercialization of non-timber forest products in agroforestry systems. Proceedings of an international conference held in Nairobi. February 19-23, 1996, Kenya. Non-wood forest products 9. FAO, Rome, Italy. Available at: http://tinyurl.com/jad49ux.

Sène, A. and Ndione, C.M., 2007. La situation de la filière gomme arabique au Sénégal. Projet ACACIAGUM, Dakar, Senegal.

Shackleton, C.M. and Shackleton, S.E., 2004. The importance of non-timber forest products in rural livelihood security and as safety nets: a review of evidence from South Africa. South African Journal of Science 100: 658-664.

Shackleton, C.M., Shackleton, S.E., Buiten, E. and Bird, N., 2007. The importance of dry woodlands and forests in rural livelihoods and poverty alleviation in South Africa. Forest Policy and Economics 9: 558-577.

Shanley, P., Pierce, A., Laird, S. and Robinson, D., 2008. Beyond timber: certification and management of non-timber forest products. CIFOR, Bogor, Indonesia.

Somo, A.A., 2009. Business plan for the purchase and export of gum Arabic from Kenya. Golden Gums K LTD, Isiolo, Kenya.

Swinnen, J.F.M. and Maertens, M., 2007. Globalization, privatization and vertical coordination in value chains in developing and transition countries. Agricultural Economics 37: 89-102.

Tollens, E.F. and Gilbert, C.L., 2003. Does market liberalisation jeopardise export quality? Cameroonian cocoa, 1988-2000. Journal of African Economies 12: 303-342.

Valdafrique, 2011: Specifications of gum Arabic. Available at: http://www.valdafrique.com.

Wekesa, C., Makenzi, P.M., Chikamai, B.N., Lelon, J.K., Luvanda, A.M. and Muga, M.O., 2010. Gum Arabic yield in different varieties. African Journal of Plant Science 3: 263-276.

Wickens, G.E., Seif El Din, A.G., Guinko, S. and Ibrahim, N., 1995. Role of acacia species in the rural economy of dry Africa and the near east. FAO conservation guide 27. FAO, Rome, Italy.

Williams, P.A. and Phillips, G.O., 2009. Gum Arabic. In: Phillips, G.O. and Williams, P.A. (eds.) Handbook of hydrocolloids. Woodhead Publishing Ltd, Cambridge, UK, pp 252-273.

Appendix 11.1. Field assessment of quality and results of chemical analyses for 27 gum arabic samples.

Village	Sample no.	Field assessment			Laboratory assessment[1]		
		Sample rank	Description		DM (%)	MM (%)	SRP (rotation angle)
			Size of nodules	Cleanliness			
L1	CEH1	1st best	large pieces	clean	88.76	4.30	-28.11
L1	CEH2	1st	small pieces	clean	89.33	3.49	-22.03
L1	CEH3	1st	small pieces	clean	88.92	9.71	-30.26
L1	CEH4	2nd	small pieces	dirty	89.89	7.55	-29.17
G1	CEH5	1st	large pieces	clean	89.10	3.65	-31.48
G1	CEH6	2nd	large pieces	dirty	89.94	16.58	-32.33
G1	CEH7	standard	large pieces	a bit clean	89.49	4.74	-29.41
Y	CEH8	1st	large pieces	a bit clean	88.97	3.99	-32.35
Y	CEH9	2nd	large pieces	dirty	90.00	6.78	-30.50
V	CEH10	1st best	large pieces	clean	89.26	3.66	-29.79
V	CEH11	1st	large pieces	dirty	88.55	3.44	-32.27
V	CEH12	2nd	small pieces	very dirty	88.59	3.81	-29.54
T	CEH13	1st	large pieces	clean	88.63	3.39	-25.04
T	CEH14	2nd	large pieces with debris	clean	88.90	3.89	-29.21
T	CEH15	standard	large pieces with debris	clean	88.73	4.14	-31.70
L2	CEH16	1st	large	dirty	89.30	3.37	-29.62
L2	CEH17	2nd	large	dirty	89.42	9.69	-18.96
L2	CEH18	3rd	large	dirty	88.48	14.47	-28.85
S1	CEH19	standard	small pieces	clean	88.08	3.01	-30.36
D	CEH20	standard	small pieces	clean	88.33	2.72	-34.11
G2	CEH21	standard	very small pieces	clean	88.53	3.19	-31.86
S2	CEH22	1st	small pieces	clean	88.11	8.07	-29.84
S2	CEH23	2nd	small pieces	clean	88.44	8.22	-27.48
S2	CEH24	3rd	small pieces with debris	quite clean	88.40	3.70	-32.19
K	CEH25	1st	very small pieces	dirty	88.69	4.82	-36.96
K	CEH26	2nd	very small pieces with debris	very dirty	88.90	4.77	-32.66
K	CEH27	3rd	very small pieces with debris	most dirty	88.63	5.50	-37.30

[1] DM = dry matter; MM = mineral matter; SRP = specific rotatory power; MM and SRP are results on DM basis.

12. Co-innovation for quality in African food chains: discovering integrated quality solutions

A. Groot Kormelinck and J. Bijman*

Wageningen University, Management Studies Group, Hollandseweg 1, 6706 KN Wageningen, the Netherlands; annemarie.grootkormelinck@wur.nl

Abstract

Quality has become more important in African food chains. Integrated quality solutions are needed to ensure that chain actors comply with increasingly stringent food safety and quality standards. This concluding chapter synthesises the main results of the previous book chapters. Food quality cannot be improved by a focus on technical solutions only. Different organisational arrangements, such as partnerships, contract farming and producer organisations, can coordinate quality requirements among chain actors. In addition, an enabling institutional environment is required. The technical, organisational and institutional elements should be combined in a co-innovation approach, which demands collaboration from multiple actors at different parts of the value chain. Co-innovation processes for quality improvements can only be built gradually and demands on-the-ground experimentation. Finally, explicit efforts are needed to guarantee sustained inclusion of smallholder farmers in higher-quality markets.

Keywords: integrated quality solutions, co-innovation, value chains, developing countries

12.1 Introduction

This book has been a search for integrated solutions for quality improvement in African food value chains. Quality is playing an increasingly important role in African food value chains. Drivers like globalisation of food systems, the spread of new technologies, urbanisation and income growth have led to a modernisation of African food value chains. This modernisation entails the existence of new value chains and the emergence of new players, such as large multinational food companies and (inter) national supermarkets. It has also led to increasingly stringent food quality standards and a rise of private quality standards (Reardon *et al.*, 2009).

The increasingly stringent criteria for quality in African food value chains presents both challenges and opportunities for smallholder farmers. Market access for smallholder farmers is considered a key development priority by international donor organisations to improve smallholder livelihoods, and to fulfil a key role in feeding the growing world population. Nonetheless, if smallholders are expected to take part of and profit from modern market participation, they need to have the resources, capabilities and surrounding enabling environment to meet quality requirements of higher-quality markets – which is still often a key challenge.

In the search for integrated solutions for quality improvement in food value chains, three perspectives should be placed central. First, an *interdisciplinary perspective* helps to understand how improvements in technology need to be aligned with changes in organisational and institutional support for those changes. Second, a *food chain perspective* shows that quality improvement depends on the coordinated activities of different actors producing and handling food products in the value chain. Third, an *innovation perspective* makes researchers and managers acknowledge that quality improvement is a process of innovation, with innovation being the result of the interplay of many actors and processes.

In this final chapter, we conclude on our discovery of integrated quality solutions. The chapter synthesises the results from the preceding eleven chapters into five conclusions. The five conclusions are discussed in separate sections, with each section consisting of an introduction, example and conclusion part. The chapter ends with directions for further research.

12.2 Beyond technical solutions

12.2.1 Introduction

The first conclusion is that food quality cannot be improved by focusing on technical solutions only. Without a doubt, technical quality solutions are needed for improving quality in value chains. Technical solutions, such as improved crop varieties, or improvements in cultivation, storage, transport, processing or packaging techniques, play a key role for improving the quality of food products in different steps of the value chain. Nevertheless, we argue that because of the complexity of the concept of food quality, there is the need to go broader than is possible by just looking at technical solutions.

Food quality in food value chains is an elusive concept for three main reasons. First, quality is defined differently at different stages of the value chain. Where producers would focus on aspects of yield, disease tolerance and uniformity, wholesalers and retailers concentrate on aspects of shelf life and availability, and consumers look at aspects of nutritional value and taste (Ruben *et al.*, 2007).

Second, there is a dichotomy in quality-attributes. This dichotomy exists between intrinsic, product-related attributes (such as size and colour) and extrinsic, process-related attributes (such as taste, appearance and nutritional value). Hence quality can refer both to the product itself, as well as to the process conditions under which it is produced, transported, packaged and sold. This dichotomy sharpens the divergence in quality perception among the different value chain actors (Ponte and Gibbon, 2005).

Third, quality requirements differ considerably between markets. Jaffee *et al.* (2011) distinguish six markets, each with a different level of quality requirements. This ranges from local (wet) markets with few quality requirements, to local supermarkets with modest quality requirements, to foreign supermarkets that have the highest quality requirements. Hence, upgrading to higher quality markets entails higher levels of quality requirements, resulting in higher compliance costs and transaction risks for producers and other actors.

Below we elaborate on three different examples indicating that a focus on technical solutions falls short for achieving the desired improvements for reaching high-quality value chains.

12.2.2 Examples

Chapter 11 by Muajawamariya *et al.* on quality management of Arabic gum in Senegal indicates that there is a large gap between quality perceptions on the production and consumption side, especially in the highest-demanding export markets. Arabic gum is not directly consumed, but is processed into different food and non-food products, which makes only intrinsic gum quality attributes important (e.g. protein content, molecular mass). However, these intrinsic attributes are not observable to farmers and collectors, who select gum based on extrinsic attributes (e.g. size and cleanliness of modules). A mismatch arises, with high quality perceived gum as perceived by farmers and collectors being assessed as low quality in the laboratories. This mismatch influences rejection rates and brings economic damage to producers and traders. The problem is further aggravated by the absence of an official grading system. Hence, the gum quality as such is not the key problem in this case. Rather, it is the incapability of traders to purchase gum with quality levels that are consistent with quality requirements of processors and consumers.

Chapters 3 and 4 by Arinloye *et al.* on quality challenges in the Beninese pineapple chain show how a policy measure focused on technical upgrading was insufficient for reaching desired quality levels. These chapters describe how the Beninese government supports technical training on pineapple juice processing to new processors in order to facilitate product and market differentiation. Yet, despite expanding production, processors find it difficult to source pineapples with a consistent quality to produce high-quality juice. Key weaknesses are that the initiative was designed as a short-term project, that linkages to and collaboration with other chain actors were not sought, and that there was a general lack of support for smallholder farmers to access targeted pineapple markets. Additional organisational and institutional efforts were needed to address poor quality coordination mechanisms and overcome information asymmetry of chain actors (see also Bitzer and Bijman, 2015; Hounhouigan, 2014).

Chapters 5 and 6 by Hirpa *et al.* and Abebe *et al.* reveal that there were mostly non-technical reasons behind a limited uptake of improved seed and ware potatoes in Ethiopia. These chapters show that adoption of improved varieties (a technical innovation) remained low. The low adoption rate was caused by a lack of information among ware growers about the availability of varieties and on the production and market conditions of improved varieties. This was related to individual adoption decisions among ware growers and was aggravated by the lack of an adequate extension system. Moreover, no attention had been given to market demands of potato varieties by different value chain actors, and the new potato varieties were developed without the involvement of ware potato producers and other value chain actors. Hence the modern varieties were not developed in a coordinated chain effort (see also Abebe *et al.*, 2013).

12.2.3 Conclusion

The three examples show that technical solutions alone are insufficient to meet standards of higher-quality value chains. We make two additional observations. First, the examples underlie the complexity of the process of quality improvement. Results from the book chapters reveal that quality definitions differ across value chain actors. The examples seem to confirm the dichotomy between intrinsic (product-related attributes) and extrinsic (process-related attributes). The examples of Ethiopia and Benin have shown that quality requirements differ significantly for different market outlets.

Second, the three examples indicate that there exists misalignment of quality requirements between different value chain actors. The examples have shown that quality preferences and requirements are weakly aligned between different actors of the value chain. In the examples, producers focused mostly on intrinsic, product-related attributes, whereas collectors, processors, retailers and other downstream actors focused more on extrinsic, process-related attributes. In all three cases, this misalignment occurs at the more upstream parts of the food value chain, between collectors and processors (gum in Senegal), between producers and processors (pineapple in Benin), and between seed and ware potato producers (potato in Ethiopia). These observations are in line with quality conceptualisations of Ponte and Gibbon (2005), Ruben *et al.* (2007), and Jaffee *et al.* (2011).

12.3 Coordination via organisational arrangements[1]

12.3.1 Introduction

The second conclusion is that quality improvement needs coordination via organisational arrangements. This builds on the first conclusion which revealed the lack

---

[1] Chapter 2 uses the term institutional arrangements, in line with the literature in New Institutional Economics (particularly North, 1990). However, we prefer to use organisational arrangements, as that makes clear that there are real organisations involved.

of quality coordination between value chain actors. Organisational arrangements can facilitate information exchange about quality requirements, strengthen coordination among actors, encourage quality-related investments and ensure greater control over product quality and safety. In doing so, these arrangements can reduce transaction costs, information asymmetry and can help to overcome key market and institutional failures (Dorward *et al.*, 2005; Royer and Bijman, 2012). This reduction is especially important in higher-quality value chains, as these have more intermediary actors and more stringent food safety and traceability requirements (Mano *et al.*, 2011; Suzuki *et al.*, 2011).

We distinguish between internal and external organisational arrangements. Two internal organisational arrangements are producer organisations and contract farming. Both arrangements consist of chain actors (producers, agribusiness firms) who are involved in bilateral or multilateral transactions with products in the value chain. External organisational arrangements, such as public-private partnerships, consist of both chain enablers and chain actors, and are often coordinated by a chain enabler. Chain enablers do not operate directly in the value chain, but operate in the wider institutional environment and fulfil key facilitating roles for the value chain to function properly. Examples are NGOs, extension services, government agencies, banks and input suppliers.

Internal and external arrangements fulfil different yet often complementary roles in quality improvement. Contract farming can assist smallholders in the provision of market linkages, inputs and extension services, whereas producer organisations are more useful when the provision of inputs, bargaining power, and market access are lacking, and partnerships, when the provision of extension services is lacking. However, these arrangements can also work complementary by utilising their comparative advantages. For example, producer organisations can support the efficiency and equity of contract farming, and partnerships can complement producer organisations in providing financial and technical assistance.

A key question is not whether, but rather *how* different internal and external organisational arrangements play a role in improving quality in value chains and in linking farmers to higher-quality markets. We now turn to several examples from our book on this issue.

12.3.2 Examples

Chapter 7 by Bitzer and Obi discusses the role of strategic partnerships in the South African citrus sector. Since the mid-2000s, the South African government promotes such partnerships as a new form of organisational arrangement. The partnerships aim to connect smallholder farmer access to higher-quality markets by providing farmer support and working on cooperation among value chain actors, particularly between established agribusinesses and emerging farmers. A first evaluation on their effectiveness reveals that these partnerships appear to be better than old policy stimuli (see also Bitzer and Bijman, 2015). The partnerships manage to promote product and process upgrading. However, the partnerships are unable to close the gap between emerging and commercial farmers. Due to the powerful position of agribusinesses who act as double gatekeepers, further upgrading of smallholder farmers is hampered. So whereas the partnerships improve smallholder market access to some extent, they may serve agribusinesses' objectives more than they serve emerging farmers need. The partnerships seem to fail to promote the transformation of emerging farmers into viable and independent farmers.

Chapter 9 by Helmsing and Enzama compares organisational arrangements between asparagus and honey value chains in respectively Peru and Uganda. The case from Peru shows that the successes from the organisational contracting arrangement were dependent on the efforts from enabling organisations active in the wider institutional environment. The NGO soon found out that the initial contracting arrangement was not successful. The NGO adapted the arrangement to changes in the broader institutional environment, especially to the new agricultural policy of the Peruvian government, and to long-term financial support by multiple (inter)national donors. In addition, the NGO was also able to respond to different endogenous and exogenous shocks, such as countering the export strategy of China and overcoming climatic shocks from el Niño. Hence, the organisational arrangement only became successful when it responded to both positive and negative events arising from the wider institutional environment.

The honey case from Uganda in Chapter 9 shows how a firm in a contract farming arrangement made insufficient use of support from indirect organisational institutions from the enabling environment. This ultimately led to the firm's collapse. The firm tried to make use of other organisations, by building on national agricultural policies and by collaborating with an NGO and the national agricultural advisory services. However, the NGO and advisory services lacked specific bee-keeping expertise. Consequently, the firm started to engage in all the activities itself and quickly got overwhelmed. With the firm taking more roles than anticipated, the interplay of contract farming with producer organisations became more complicated and skewed along the way. As a result, the value chain became more complex, which led to the erosion of a former success story.

Chapter 10 by Adekambi *et al.* illustrates the role of informal organisational arrangements in shrimp value chains in Benin. Shrimp fishers use their social networks to establish informal relations with local traders. These fishers face a number of limitations, such as a lack of market information, limited contact with extension services, a weak basic infrastructure and legal constraints. These shortcomings discourage fishers to improve quality and restrict their access to export markets due to high transaction costs. As a result, traders start to share market information with the shrimp fishers they trust. Traders provide these fishers with a loan, train them in specific fish handling skills, and offer fixed purchasing pries. Moreover, traders enforce quality standards upon the fishers, thus providing better guarantees of compliance with the stringent requirement of export markets. Hence, fishers and traders generally rely on each other's trustworthiness to run and sustain their businesses. This enables fishers to take effective fishing decisions to improve quality and meet the expectations of customers.

12.3.3 Conclusion

Our conclusion that quality improvement needs coordination via organisational arrangements is based on three observations. First, internal organisational arrangements should better align their mutual activities as to ease value chain coordination. A suboptimal collaboration of internal arrangements, such as contract farming with producer organisations, may enlarge the institutional gaps instead of reduce them (Suzuki *et al.*, 2011). The Uganda case from Chapter 9 showed this, where problems between the firm and producer organisation escalated as tasks and activities were unequally divided between them, resulting in amplified information asymmetry and transaction costs. Contracting firms need to coordinate quality requirements of downstream chain actors so that producer organisations become tailored to comply with these requirements. Producer organisations should change their internal organisational functioning to control free-riding, improve power balances and enhance mutual quality control (Narrod *et al.*, 2009; Van Tilburg *et al.*, 2007).

Second, internal organisational arrangements could be combined with external organisational arrangements. At one hand, internal arrangements alone cannot improve quality without wider institutional support. The examples from Chapter 9 have shown that success can be achieved when the internal arrangements are able to react and build on external institutional support (Peru), or lead to collapsing value chains when this is not sufficiently done (Uganda). On the other hand, external arrangements could cooperate with internal arrangements. The example from Chapter 7 has shown that partnerships alone were insufficiently able to coordinate quality. As the partnership consisted merely of chain enablers, it stood too much on the side and had too little influence on powerful chain actors. Partnerships can collaborate with producer organisations to guarantee that smallholder producers are not excluded from contracting arrangements. Especially in the initial phase of setting up contract farming, partnerships can collaborate with producer organisations to address the lack of extension services, quality control, technical assistance and training (Boselie *et al.*, 2003; Narrod *et al.*, 2009).

Third, informal arrangements can also be effective in quality promotion. Informal arrangements, such as farmer groups or trader networks, can reduce transaction costs by transmitting quality requirements, ensuring collective action among producers and establishing informal trade relations (Granovetter, 1985). For instance, Chapter 10 has shown that informal trade relations between shrimp fishers and traders in Benin may be a productive strategy in enhancing shrimp fishers' integration with export chains. Social networks of fishers, which are governed by trust and reputation, can safeguard against opportunistic behaviour and can strengthen their position in the value chain (Gulati *et al.,* 2007; Saenger *et al.,* 2014). Informal arrangements are especially important upstream in value chains and in more remote rural areas (Burgess and Steenkamp, 2006).

Concluding, our results stress the importance of combining different organisational arrangements to overcome constraints and achieve quality improvements. This also requires the embeddedness in a conducive wider institutional environment, which leads to the next conclusion.

12.4 The wider institutional environment

12.4.1 Introduction

The third conclusion is that a conducive institutional environment is needed for quality improvement. Organisational arrangements alone, or in combination with each other, are insufficient for sustainably linking farmers to higher quality markets. These organisational arrangements also need to be embedded in a well-functioning institutional environment.

The institutional environment refers to the institutions that frame the economic activity by setting mutual social expectations. These institutions can be either formal (laws, public norms, property rights, education systems) or informal (social norms of conduct, levels of trust, customs, community unwritten rules) (North, 1990). The institutional environment may have a direct or an indirect impact on quality improvement. In many developing countries, a weak institutional environment entails low information availability, high coordination costs and high risks, as well as poor infrastructure in terms of roads, transport, storage and telecommunication networks (World Bank, 2007).

We emphasise that the institutional environment changes very slowly over time. Williamson (2000) distinguishes four types of institutions based on their time span of change. Following informal institutions that take the longest time to change (once in hundred to a thousand years), the institutional environment changes only once every ten to hundred years. Successively, governance structures, such as contract farming, partnerships and producer organisations, change more rapidly, that is one to ten years. Ultimately, resource allocation within these governance structures changes continuously.

The ability of the institutional environment to change is increasingly being tested. The institutional environment has to respond to endogenous or exogenous shocks and to the introduction of new technologies (Hodgson, 2006; Nelson, 2008). In food value chains, different upgrading strategies are adopted to add quality and thus value to products. Zooming in on these upgrading strategies, four categories can be distinguished: (1) process upgrading, that is reorganising production processes for greater efficiency and quality; (2) product upgrading, thus more sophisticated products; (3) functional upgrading, or acquiring new functions to increase the skills needed for activities; and (4) inter-sectoral upgrading, which means moving into new productive activities (Giuliani *et al.*, 2005; Humphrey and Schmitz, 2002).

Upgrading is often coupled with an effort to climb the 'ladder of market access', being the diversification into higher-value markets (Jaffee *et al.*, 2011). Upgrading opportunities can be facilitated by both organisational arrangements, for instance promoting certain types of upgrading through contract farming or a partnership, and by the wider institutional environment, for instance through public policies (Pietrobelli, 2008). Nonetheless, moving up the ladder of market access involves an increase in quality requirements, implying higher compliance costs and transaction risks for producers and other chain actors. This makes upgrading to higher-quality markets a difficult process, whereby success is not guaranteed (Gulati *et al.*, 2007).

Whereas upgrading strategies are often implemented at a rapid pace, developing an institutional environment takes a long time. This raises questions on the interplay between quality upgrading strategies, endogenous and exogenous shocks and the response of the institutional environment. We will turn to three book examples on this topic.

12.4.2 Examples

Chapter 7 by Bitzer and Obi describes how a sudden privatisation led to an institutional void in the South African citrus sector. In 1997, the countries' citrus sector drastically changed from a publicly regulated single-marketing channel system into a liberal market system and the entrance of new exporter companies. In the years after the deregulation, quality control became a tremendous challenge because vital institutional services to citrus producers were eliminated, such as research support, the quality control board, extension, and information dissemination services. Subsequently, the citrus sector became in a deep crisis, resulting in plummeting export earnings and liquidated farms.

Chapter 9 by Helmsing and Enzama describes how the upgrading strategy of a lead firm in the Ugandan honey value chain led to the collapse of the value chain. The firm established and controlled the entire value chain itself, and immediately targeted the highest-quality export value chains through supplying certified organic honey products. This rapid upgrading, in combination with the limited use of the wider institutional support, proved to be too much too soon. The firm could no longer maintain its monopsonist function, went bankrupt, and the value chain started eroding. Interestingly, other companies noticed the efforts already done by the firm and started to engage in the sector. Former firm staff started operating as traders, and NGOs, local governments and private enterprises started hiring private service providers to introduce alternative technologies and to contract producers.

Chapter 10 by Adekambi *et al.* unravels how an upgrading strategy led to targeting the highest market ladder in the shrimp sector in Benin. From 2006 onwards, the Beninese government targeted export market integration for shrimps. However, the export chain made a false start because of a mismatch between the safety levels required by the EU and the quality levels that Beninese chain actors could deliver. The inconsistency between the provisions of the laws and their implementing decrees has been a main weakness of the food safety law in Benin. Consequently, the sector had been in decline for over ten years due to difficulties of complying with EU requirements. On hindsight, the food safety and quality problems should probably have been anticipated when the chain was created in the 1990s. Repairing the system afterwards appeared a painful process in which much of the prior investments were lost.

12.4.3 Conclusion

Because a conducive institutional environment is needed for quality upgrading, processes of quality improvement cannot be successful overnight. Targeting higher-value markets can only be done gradually if the enabling institutional environment is not in place yet. All three examples reveal how a sudden upgrading to the highest quality export markets – thereby skipping several ladders of market access – led to huge crises in the respective agriculture and fisheries sectors. The upgrading went entirely wrong, as it was done at such a rapid pace that it became impossible for the institutional environment to respond timely.

The examples also made clear that it took the respective sectors many years to overcome its crises. Repairing the damage and gradually returning back to higher quality markets proved to be a much longer road than would have been the case with slowly upgrading. This relates to the work of Brousseau and Raynaud (2011) who advocate that institutional development is a lengthy and path-dependent process with different institutions being hierarchical and interrelated. Our conclusion also aligns with Williamson's (2000) typology of gradual change among institutions that showed the long duration of changing the institutional environment.

12.5 Co-innovation

12.5.1 Introduction

On the basis of the three conclusions presented above, we can draw a fourth conclusion, namely that quality improvement in value chains can only be successful if it follows a co-innovation approach. The first conclusion showed how technological solutions alone are insufficient to come to higher quality food chains because quality improvement is about coordination of quality requirements among chain actors. This led to the second conclusion, stating that such coordination should take place via internal and external organisational arrangements, such as contract farming, producer organisations and partnerships. The third conclusion showed how these organisational arrangements alone are insufficient to improve quality as they require a conducive institutional environment.

Co-innovation implies the integration of these technical, organisational and institutional elements. Our co-innovation approach outlined here builds on the exploration of the co-innovation concept by Bitzer and Bijman (2015). Co-innovation is a relatively new concept from management thinking on innovation, which builds upon the integration of recent advances in innovation system research and value chain literature in developing country agriculture. We elaborate on the three dimensions of co-innovation and apply these to our book chapters.

12.5.2 Examples

First, co-innovation stands for collaborative innovation as a multi-actor process that combines different types of knowledge and resources from different actors. Chapter 7 has shown how partnerships in the citrus sector in South Africa operated too much from the institutional environment and therefore had limited direct influence on chain actors, especially on powerful agribusinesses. The opposite occurred in the examples from Peru and Uganda in Chapter 9, where chain actors insufficiently sought support from other organisations. As a result, both in Chapter 7 and 9, suboptimal quality solutions were reached.

Second, co-innovation demands a multi-dimensional process in which changes in technology, organisation and institutions are effectively combined. This dimension is clearly emerging from our first three conclusions. Chapter 3 and 4 have shown that a strategy by the Beninese government for quality improvement in the pineapple chain was too much focused on a technical solution (training pineapple processors). No linkage was sought to organisational solutions, such as linking farmers and other value chain actors, and institutional solutions for technical services. For that reason, the innovation had limited success.

Third, co-innovation means coordinating innovations at different levels of the value chain. Chapter 11 has revealed how quality requirements of Arabic gum in Senegal were not coordinated among chain actors, leading to divergent quality perceptions between upstream and downstream parts of the value chain. Similarly, Chapters 5 and 6 have shown how preferences for high quality potatoes in Ethiopia were not aligned between seed and ware potato growers upstream in the chain and between other chain actors further downstream in the value chain. This led to a mismatch in the supply and demand of improved varieties.

An example of co-innovation is the following case about mobile phone technologies in the Beninese pineapple chain. Chapter 3 by Arinloye *et al.* shows that key institutional constraints for improving quality and reaching higher-value chains are the lack of market information, a weak physical infrastructure, particularly roads, transport and storage facilities, and a weak information structure, particularly limited information exchange on quality among chain actors. Chapter 4 shows how a technological innovation, a mobile phone service that provides up-to-date market information, can help farmers in choosing a market outlet and assist in bargaining with traders. Farmers are willing to pay a price for product price and quality information. Especially farmers far from the urban centre who trade mostly with local buyers, and farmers with limited contact with agricultural extension services are keen to pay for this service (see also Arinloye *et al.*, 2015).

In this mobile-phone example from Benin, all three elements of co-innovation can be found. First, engagement from multiple actors is required. For instance, a private mobile phone operator is needed, and public actors, such as national statistical and market information institutes, have to collect the market and price information that feed into the mobile service. Second, technical, organisational and institutional elements need to be combined. The technical innovation of the mobile-based service cannot function without a physical and digital institutional infrastructure, such as functioning ICT networks, legislation and policies. The innovation also demands organisational support from public/private actors who, for instance, financially support the set-up of the service, and who communicate the service to farmers. Third, the innovation is used and supported by actors on different actors in and surrounding the value chain, from local farmers to NGOs, research institutes and mobile providers.

12.5.3 Conclusion

The conclusion that quality improvement in value chains needs co-innovation comes with two additional statements. First, co-innovation can only be sustainable when all three elements are fulfilled and when they are combined simultaneously and complementary. The mobile phone example has shown that when the three dimensions of co-innovation converge, key institutional constraints to quality improvements in chains can be resolved. Examples from the first three conclusions have shown that either one of the elements of co-innovation was missing, or that the links between the elements were weak. Following Williamson (2000) on time horizons of institutional change, we emphasise that sustainable co-innovation should take into account that elements of co-innovation change at a different pace. Technical innovations come sudden, the organisational arrangements need a longer time, followed by the institutional environment, which takes the longest time to adapt to the innovation.

Second, co-innovation cannot be designed upfront, but needs considerable on-the-ground experimentation. The Peru case described in Chapter 9.2.3 shows that quality upgrading only became successful when the lead chain actor started to adapt the contracting arrangement based on ongoing events and shocks present in the wider institutional environment. This and other examples show that successful co-innovation only occurs when there is sufficient experimentation between technical, organisational and institutional changes. The argument is supported by reasoning from New Institutional Economics (NIE), which states that actors are faced by situated bounded rationality because of which they cannot see threats to co-innovation at forehand. Therefore, for co-innovation to be sustainable, it cannot be designed externally nor at forehead. Rather, co-innovation needs some kind of decentralised experimentation to ensure that all complementary and co-evolving changes match (Williamson, 1975, 2000).

The elaboration of co-innovation bring us to the question of how to develop sustainable co-innovation that is inclusive to smallholder farmers. This is discussed in the next section.

12.6 Benefits for smallholder participation

12.6.1 Introduction

The fifth conclusion advocates that specific attention is required to guarantee smallholder farmer inclusion in higher-quality value chains and to ensure long-term benefits from this participation. Upgrading to higher-quality markets is still a major problem for many smallholders. Farmers face different institutional constraints that limit their opportunities for quality upgrading (Maertens and Swinnen, 2009). The topic of 'linking farmers to markets' has become an important topic on the international agendas for contributing to food security and rural development. Three key themes seem to dominate the topic.

First is the debate on whether smallholder farmers are at a competitive disadvantage *vis-a-vis* large farmers for inclusion in higher-quality markets. Whereas most literature seems to indicate that smallholders are at a comparative disadvantage as compared to large farms, their exclusion from higher-quality chains is not necessarily the case. Large farms usually profit from higher economies of scale and lower unit transaction costs. Smallholder farms, nonetheless, are more flexible in applying family labour and can build on their intensive local knowledge (Poulton *et al.*, 2010). In addition, small farms usually have a higher land productivity, while large farms have a higher labour productivity. Thus, in situations with abundant labour and scarce land resources, small farms are more efficient. Still, farm size may not always be the most important determinant for inclusion of farmers, but instead it could be farms' location to roads and processing plants, or farmers' managerial ability that leads to inclusion or exclusion (Da Silva and Rankin, 2013).

A second debate concerns the sustainability of higher-quality market participation. Both inclusion in and exclusion from higher-value chains are often not permanent, as farmers can move to higher or lower-quality chains in the ladder of market access. A sustainable inclusion of farmers is to a large extent determined by their relations with downstream buyers, for example with firms in contract arrangements. Firms substitute farmers when they can reduce costs or are dissatisfied with current suppliers. Empirical contract farming studies show high farmer exit rates, indicating that farmer experiences and their benefits might be variable (Barrett *et al.*, 2012; Narayanan, 2014).

A third debate is about the impacts of inclusion of farmers in high-quality chains. Positive impacts on incomes from high-value market participation by smallholders are reported by for example Birthal *et al.* (2008), Maertens and Swinnen (2009), and Minten *et al.* (2009). Even though there exist critical reports, most of the evidence suggests that, under appropriate enabling environments, the potential advantages of contracting for farmers and agribusiness firms tend to outweigh the potential disadvantages (Da Silva and Rankin, 2013; Prowse, 2012).

Zooming in on farmers' access to higher quality markets necessitates to investigate the type of resources and capacities that are at farmers' disposal. Chapter 7 and 8 discuss the topic of farmers resources. We conclude with three core debates in the 'farmers to markets' topic.

12.6.2 Examples

Chapter 8 by Grwambi *et al.* on citrus fruits in South Africa focuses on the types of resources that smallholder farmers need in order to increase competitiveness and access higher value markets. Such types of resources are production factors (land, labour, capital and water); personal skills and knowledge; process capabilities (logistic, marketing and planning expertise); and dynamic capabilities, i.e. the ability to make strategic changes in the resource base to keep up with changing market environments. Case comparisons disclose how the nature of resources at smallholder disposal is linked to their potential to add value to the product and hence to have the capacity to participate in high value markets. The chapter shows that especially farmers that do not have the resources to add value to the product produce lower quality, and are more likely to cater lower quality markets. Especially process and dynamic capabilities are key for targeting high value markets.

Chapter 7 by Bitzer and Obi, also on citrus fruit in South Africa, shows how the sudden privatisation of the citrus sector and the accompanying rise of private quality standards led to exclusion of smallholder farmers. The privatisation and accompanying rise of quality standards put enormous pressure on farmers and other chain actors to make the transition towards these higher-quality export chains. Assistance from a single channel exporter was withdrawn and was not replaced by adequate public support, implying a shortage of essential smallholder services. Those farmers wishing to remain competitive had to implement alternative upgrading strategies. Yet not all farmers were able to do such upgrading. Whereas larger farmers mostly managed to address these challenges by means of significant upgrading activities, smallholder farmers largely failed to climb the 'ladder of market access', and ended up confined to the informal domestic market. Especially smallholder black farmers were excluded, reinforcing the historical exclusion of smallholder farmers from remunerative markets. This exclusion added to the structural duality of South-African agriculture that runs along racial lines.

12.6.3 Conclusion

The fifth conclusion on ensuring smallholder inclusion in higher-quality value chains comes with two remarks. First, farmers need to have access to different types of resources and be part of a conducive co-innovation environment to be included in higher-quality value chains. Chapter 8 has shown that farmers' ability to upgrade and be included in higher-quality markets is based on their capacities and resources. Chapter 7 has indicated that both the capacities and resources of farmers and the support from the organisational and institutional environment determines in- or exclusion. The two examples show that, to induce smallholders to increase production, a combination of pull and push factors needs to be in place. Contributing to the inclusion debate, Chapter 7 has revealed that mostly smallholder producers were excluded from higher-quality chains. This is in line with results of, for instance, Da Silva (2005), Kirsten and Sartorius (2002), and Singh (2002); but is contradicting more positive inclusion studies of Swinnen and Maertens (2007) and Reardon *et al.* (2009).

Second, even if farmers are included in higher-quality chains, it is not guaranteed that they are sustainably included and obtain long-term benefits of participation. Chapter 7 shows that inclusion is not necessarily permanent, as an external shock (e.g. privatisation of the sector) can lead to sudden exclusion of farmers from high-quality chains. The reverse can also occur in which farmers upgrade to higher quality markets through, for example, upgrading their production processes (Perez-Aleman, 2011). An additional question relates to which markets are best suited for different types of farmers. As not all farmers have the resources and skills and institutional support to produce for the highest quality markets, an increased differentiation may be seen in the development of food chains. Farmers and other chain actors, as well as the institutional environment, need to provide sustainable support and prove they are capable of longer-term thinking and acting.

12.7 Further research

Based on the various studies presented in this volume, we draw the following five suggestions for further research. These five recommendations can be directly linked to the five conclusions.

First, a better understanding is needed on how the introduction of new technologies leads to required changes in organisational arrangements and the institutional environment. There is augmented policy attention for ICT technologies that are increasingly applied in the agricultural and rural development domain (World Bank, 2016). Nonetheless, little is known about how such technologies change the interplay between the three components of co-innovation. For instance, the introduction of mobile technologies to farmers might reduce the need for collective action in producer organisations, because it may no longer be necessary for producer organisations to jointly collect and disseminate market information (Da Silva and Rankin, 2013). In addition, the institutional environment needs to focus on both digital and physical infrastructures as to support the technology uptake, which might demand collaboration between new and existing organisations and institutions (Roberts and Grover, 2012; World Bank, 2016). Hence, we need to know better how co-innovation can support integrated quality solutions that successfully respond to new technologies.

Second, more insights are needed on the benefits of combining different organisational arrangements. Various internal and external organisational arrangements, such as partnerships, producer organisations and contract farming, complement each other in activities and competences. Jointly they might reduce the drawbacks of individual arrangements. For instance, we need to know better how producer organisations and partnerships can cooperate to mitigate the risks of contract farming (Royer and Bijman, 2012). Besides, more insights are needed on the functioning and effectiveness of informal arrangements, such as informal farmer-trader relations, and their interaction with formal organisational arrangements (Saenger et al., 2014; Trienekens, 2011). As a result, a deepened understanding of the interactions between organisational arrangements should lead to insights on how and under what conditions combinations of different arrangements are effective for quality improvement.

Third, research needs to better include an examination of current and past socio-cultural, religious and political events that influence the co-innovation process. For instance, demand for pineapples in Benin (Chapter 3 and 4) is strongly influenced by the seasonal peak during the Ramadan, in Benin and neighbouring countries. Also, the current South African agricultural sector (and wider society) is strongly influenced by the past racial segregation. Current polices are designed to overcome the structural duality and specifically target black smallholder inclusion (Chapter 7 and 8). Hence, co-innovation is determined by past socio-cultural, religious and political factors, which is often neglected in research on quality and smallholder inclusion in value chains (Lee et al., 2012; World Bank, 2015). Such neglect could lead to the research, design and implementation of co-innovation processes that do not fit countries' agricultural sectors and thus will have limited success.

Fourth, a new research agenda should arise on the sustainability of co-innovation. Different shocks test the sustainability of co-innovation, from internal (within value chains) and external shocks (in the wider environment) to gradual (e.g. climate change, urbanisation) and unexpected shocks (natural disasters, political and financial crises). Co-innovation needs to be built gradually and should be based on on-the-ground experimentation, in order to be able to withstand and respond to these shocks. However, current studies on co-innovation for quality in food chains rarely take such sustainability aspects into account. Sustainability is needed at different components of co-innovation and also between the co-innovation components, which demands a holistic analysis of co-innovation. Interdisciplinary research should be conducted that, for instance, incorporates innovation systems and value chain analysis with the emerging field of food system resilience (Tendall et al., 2015).

Fifth, more insights are needed on the sustainability of smallholder inclusion in high-quality value chains. Smallholder participation in high-quality markets is often not sustainable, for different reasons. Farmers lack individual resources, participation in organisational arrangements such as contract farming comes at a risk, the institutional environment is not supportive, or upgrading to higher-quality chains is done too rapid. Most studies on farmers' market access go beyond the fact that it may not be farmers deciding on participation, but rather it is decided by firms or other chain actors (Herforth *et al.*, 2015). Furthermore, even if farmers have the choice, they may not prefer the most modern channel – contrary to what is often assumed. Farmers have heterogeneous preferences and might prefer attributes of traditional market outlets due to a variety of reasons (Schipmann and Qaim, 2011). Scholars, policymakers, and donors should therefore focus on the question of which markets are best suited for which types of farmers as to promote sustainable inclusion in higher-quality chains.

References

Abebe, G.K., Bijman, J., Pascucci, S. and Omta, O., 2013. Adoption of improved potato varieties in Ethiopia: the role of agricultural knowledge and innovation system and smallholder farmers' quality assessment. Agricultural Systems 122: 22-32.

Arinloye, D.D.A.A., Linnemann, A.R., Hagelaar, G., Coulibaly, O. and Omta, O.S.W.F., 2015. Taking profit from the growing use of mobile phone in benin: a contingent valuation approach for market and quality information access. Information Technology for Development 21: 44-66.

Barrett, C.B., Bachke, M.E., Bellemare, M.F., Michelson, H.C., Narayanan, S. and Walker, T.F., 2012. Smallholder Participation in contract farming: comparative evidence from five countries. World Development 40: 715-730.

Birthal, P.S., Jha, A.K., Tiongco, M.M. and Narrod, C., 2008. Improving farm to market linkages through contract farming: a case study of smallholder dairying in India. IFPRI Discussion Paper 00814. IFPRI, Washington, DC, USA.

Bitzer, V. and Bijman, J., 2015. From innovation to co-innovation? An exploration of African agrifood chains. British Food Journal 117: 2182-2199.

Boselie, D., Henson, S. and Weatherspoon, D., 2003. Supermarket procurement practices in developing countries: redefining the roles of the public and private sectors. American Journal of Agricultural Economics 85: 1155-1161.

Brousseau, E. and Raynaud, E, 2011. 'Climbing the hierarchical ladders of rules': a life-cycle theory of institutional evolution. Journal of Economic Behavior and Organization 79: 65-79.

Burgess, S.M. and Steenkamp, J.B.E.M., 2006. Marketing renaissance: how research in emerging markets advances marketing science and practice. International Journal of Research in Marketing 23: 337-356.

Da Silva, C.A., 2005. The growing role of contract farming in agri-food systems development: Drivers, theory and practice. FAO, Rome, Italy.

Da Silva, C.A. and Rankin, M. (eds.), 2013. Contract farming for inclusive market access. FAO, Rome, Italy. Available at: http://tinyurl.com/peak29m.

Dorward, A., Kydd, J., Morrison, J. and Poulton, C., 2005. Institutions, markets and economic co-ordination: linking development policy to theory and praxis. Development and Change 36: 1-25.

Giuliani, E., Pietrobelli, C. and Rabellotti, R., 2005. Upgrading in global value chains: lessons from Latin American clusters. World Development 33: 549-573.

Granovetter, M., 1985. Economic action and social structure: the problem of embeddedness. American Journal of Sociology 91: 481-510.

Gulati, A., Minot, N., Delgado, C. and Bora, S., 2007. Growth in high-value agriculture in Asia and the emergence of vertical links with farmers. Food Policy: 91-108.

Herforth, N., Theuvsen, L., Vásquez, W. and Wollni, M., 2015. Understanding participation in modern supply chains under a social network perspective – evidence from blackberry farmers in the Ecuadorian Andes. GlobalFood Discussion Papers (No. 57). Göttingen Universität, Göttingen, Germany.

Hodgson, G.M., 2006. What are institutions? Journal of Economic Issues 1: 1-25.

Hounhouigan, M.H., 2014. Quality of pasteurised pineapple juice in the context of the Beninese marketing system. PhD thesis, Wageningen University, Wageningen, the Netherlands.

Humphrey, J. and Schmitz, H., 2002. How does insertion in global value chains affect upgrading in industrial clusters? Regional Studies 36: 1017-1027.

Jaffee, S., Henson, S. and Rios, L.D., 2011. Making the grade – Smallholder farmers, emerging standards, and development assistance programs in Africa – a research program synthesis. Report No. 62324. World Bank, Washington, DC, USA.

Kirsten, J. and Sartorius, K., 2002. Linking agribusiness and small-scale farmers in developing countries: is there a new role for contract farming? Development Southern Africa 19: 503-529.

Lee, S.M., Olson, D.L. and Trimi, S., 2012. Co-innovation: convergenomics, collaboration, and co-creation for organizational values. Management Decision 50: 817-831.

Maertens, M. and Swinnen, J.F., 2009. Trade, standards, and poverty: evidence from Senegal. World Development 37: 161-178.

Mano, Y., Yamano, T., Suzuki, A. and Matsumoto, T., 2011. Local and personal networks in employment and the development of labor markets: evidence from the cut flower industry in Ethiopia. World Development 39: 1760-1770.

Minten, B., Randrianarison, L. and Swinnen, J.F.M., 2009. Global retail chains and poor farmers: evidence from Madagascar. World Development 37: 1728-1741.

Narayanan, S., 2014. Profits from participation in high value agriculture: evidence of heterogeneous benefits in contract farming schemes in Southern India. Food Policy 44: 142-157.

Narrod, C., Roy, D., Okello, J., Avendaño, B., Rich, K. and Thorat, A., 2009. Public-private partnerships and collective action in high value fruit and vegetable supply chains. Food Policy 34: 8-15.

Nelson, R.R., 2008. What enables rapid economic progress: what are the needed institutions? Research Policy 37: 1-11.

North, D.C., 1990. Institutions, institutional change and economic performance. Cambridge University Press, Cambridge, UK.

Perez-Aleman, P., 2011. Collective learning in global diffusion: spreading quality standards in a developing country cluster. Organization Science 22: 173-189.

Pietrobelli, C., 2008. Global value chains in the least developed countries of the world : threats and opportunities for local producers. International Journal Technological Learning, Innovation and Development 1: 459-481.

Ponte, S. and Gibbon, P., 2005. Quality standards, conventions and the governance of global value chains. Economy and Society 34: 1-31.

Poulton, C., Dorward, A. and Kydd, J., 2010. The future of small farms: new directions for services, institutions, and intermediation. World Development 38: 1413-1428.

Prowse, M., 2012. Contract farming in developing countries: a review. Institute of Development Policy and Management, University of Antwerp, Antwerp, Belgium.

Reardon, T., Barrett, C.B., Berdegué, J.A. and Swinnen, J.F., 2009. Agrifood industry transformation and small farmers in developing countries. World Development 37: 1717-1727.

Roberts, N. and Grover, V., 2012. Leveraging information technology infrastructure to facilitate a firm's customer agility and competitive activity: an empirical investigation. Journal of Management Information Systems 28: 231-270.

Royer, A. and Bijman, J., 2012. Towards an analytical framework linking institutions and quality: evidence from the Beninese pineapple sector. African Journal of Agricultural Research 7: 5344-5356.

Ruben, R., Van Boekel, M., Van Tilburg, A. and Trienekens, J., 2007. Governance regimes for quality management. Wageningen Academic Publishers, Wageningen, the Netherlands.

Saenger, C., Torero, M. and Qaim, M., 2014. Impact of third-party contract enforcement in agricultural markets – a field experiment in Vietnam. American Journal of Agricultural Economics 96: 1220-1238.

Schipmann, C., and Qaim, M., 2011. Supply chain differentiation, contract agriculture, and farmers' marketing preferences: the case of sweet pepper in Thailand. Food Policy 36: 667-677.

Singh, S., 2002. Contracting out solutions: political economy of contract farming in the Indian Punjab. World Development 30: 1621-1638.

Suzuki, A., Jarvis, L.S. and Sexton, R.J., 2011. Partial vertical integration, risk shifting, and product rejection in the high-value export supply chain: the Ghana pineapple sector. World Development 39: 1611-1623.

Swinnen, J.F. and Maertens, M., 2007. Globalization, privatization, and vertical coordination in food value chains in developing and transition countries. Agricultural Economics 37: 89-102.

Tendall, D.M., Joerin, J., Kopainsky, B., Edwards, P., Shreck, A., Le, Q.B. and Six, J., 2015. Food system resilience: defining the concept. Global Food Security 6: 17-23.

Trienekens, J., 2011. Agricultural value chains in developing countries a framework for analysis. International Food and Agribusiness Management Review 14: 51-82.

Van Tilburg, A., Trienekens, J., Ruben, R. and Van Boekel, M., 2007. Governance for quality management in tropical food chains. Journal on Chain and Network Science 7: 1-9.

Williamson, O.E., 1975. Markets and hierarchies: analysis and anti-trust implications: a study in the economics of internal organization. Free Press, New York, NY, USA.

Williamson, O.E., 2000. The new institutional economics – taking stock, looking ahead. Journal Economic Literature 38: 595-613.

World Bank., 2007. World Development Report 2008: agriculture for development. World bank, Washington, DC, USA.

World Bank, 2015. World Development Report 2015: mind, society and behavior. World bank, Washington, DC, USA.

World Bank., 2016. World Development Report 2016: digital dividends. World bank, Washington, DC, USA.

Printed in the United States
by Baker & Taylor Publisher Services